FLORA OF TROPICAL EAST AFRICA

MENISPERMACEAE

G. Troupin [1]

(Jardin Botanique de l'État, Brussels)

Twining or rarely erect shrubs or small trees, dioecious, with the wood in cross-section showing broad medullary rays. Leaves petiolate, sometimes peltate, without stipules, usually simple, entire or lobed. Inflorescences various, many-flowered, rarely the flowers solitary or geminate, axillary or borne on the leafless wood. Flowers small, unisexual, regular, rarely slightly irregular. Flowers ♂ : sepals 3–12 or more, rarely 1, free or slightly connate, imbricate or valvate ; petals 1–6 or absent, free or connate, usually imbricate ; stamens 3–6 or indefinite, rarely 2, free or variously united. Flowers ♀ : sepals and petals generally as in ♂ flowers, sometimes not so numerous ; staminodes absent or present ; carpels 3–6 or more, rarely 1, free ; ovules 2, soon reduced to 1 by abortion, attached to the ventral suture. Fruiting carpels drupaceous, with the scar of the style subterminal or near the base by excentric growth ; exocarp membranaceous or subcoriaceous, mesocarp more or less pulpy, endocarp often chartaceous or bony, rugose, tuberculate or ribbed. Seed often curved and horseshoe-shaped, with uniform or ruminate endosperm, or without endosperm.

Key to plants with female flowers or fruits

Leaves not peltate :
- Sepals 3–18 :
  - Inner sepals connate ; flowers axillary, more or less solitary ; carpels 4–12 . . . 1. **Epinetrum**
  - Inner sepals free :
    - Inner sepals valvate :
      - Carpels 8–40 ; sepals thick :
        - Surface of fruit glabrescent, smooth or rugose ; sepals glabrous or nearly so ; petals developed . . . . . 3. **Tiliacora**
        - Surface of fruit velutinous or puberulous ; sepals hairy ; petals minute or absent 2. **Triclisia**
      - Carpels 3–6 ; sepals thin . . . . . 5. **Dioscoreophyllum**
    - Inner sepals imbricate :
      - Flowers solitary or in fasciculate cymules ; carpels 3–6 . . . . . . 4. **Cocculus**
      - Flowers numerous, in more or less elongate and compound inflorescences :
        - Leaves deeply lobed, hirsute ; fruit bristly ; endocarp fibrillous . . 6. **Jateorhiza**

[1] The Editors are grateful to Mr. J. P. M. Brenan for translating the French text.

Leaves entire or angular :
Leaves densely hairy, especially when young ; petals with a rib on the inner side . . . . . . 7. **Chasmanthera**
Leaves entire, not angular, glabrous ; petals without a rib . . . . 8. **Tinospora**
Sepals 1, rarely 2 ; bracts enlarged in fruit . . 10. **Cissampelos**
Leaves peltate :
Bracts minute, not enlarged in fruit ; sepals 3–6 ; petals 2–4, free . . . . 9. **Stephania**
Bracts conspicuous, enlarged in fruit, membranaceous or leafy ; sepal 1 ; petal 1, rarely more. . . . . 10. **Cissampelos**

Key to plants with male flowers

Leaves not peltate :
Inner sepals connate ; stamens 15–30, united into an elongated synandrium ; flowers in shortly pedunculate cymules or glomerules . . 1. **Epinetrum**
Inner sepals free :
Petals 0 :
Stamens free ; inflorescences not elongated . 2. **Triclisia**
Stamens united into a subglobose synandrium ; inflorescences elongated . . . 5. **Dioscoreophyllum**
Petals 3–6 :
Petals more or less united into a cup ; stamens united into a stipitate synandrium ; anthers opening by a transverse slit . 10. **Cissampelos**
Petals not united into a cup :
Inner sepals valvate :
Petals developed ; inner sepals glabrous or nearly so . . . . . . 3. **Tiliacora**
Petals minute or absent . . . 2. **Triclisia**
Inner sepals imbricate :
Anthers opening by a transverse slit :
Inflorescences in long panicles of sessile flowers ; leaves large, deeply lobed, strigose . . . . . . 6. **Jateorhiza**
Inflorescences in short cymules of axillary flowers ; leaves short, generally entire, not strigose . 4. **Cocculus**
Anthers opening by a longitudinal slit :
Leaves nearly rounded or subangular, densely hairy especially when young ; petals with a rib on the inner side . . . . . . 7. **Chasmanthera**
Leaves entire, not angular ; glabrous ; petals without a rib . . . 8. **Tinospora**
Leaves peltate ; stamens always united :
Sepals 6–8 ; petals 3–4, free . . . . . 9. **Stephania**
Sepals 4 ; petals united into a cup . . . 10. **Cissampelos**

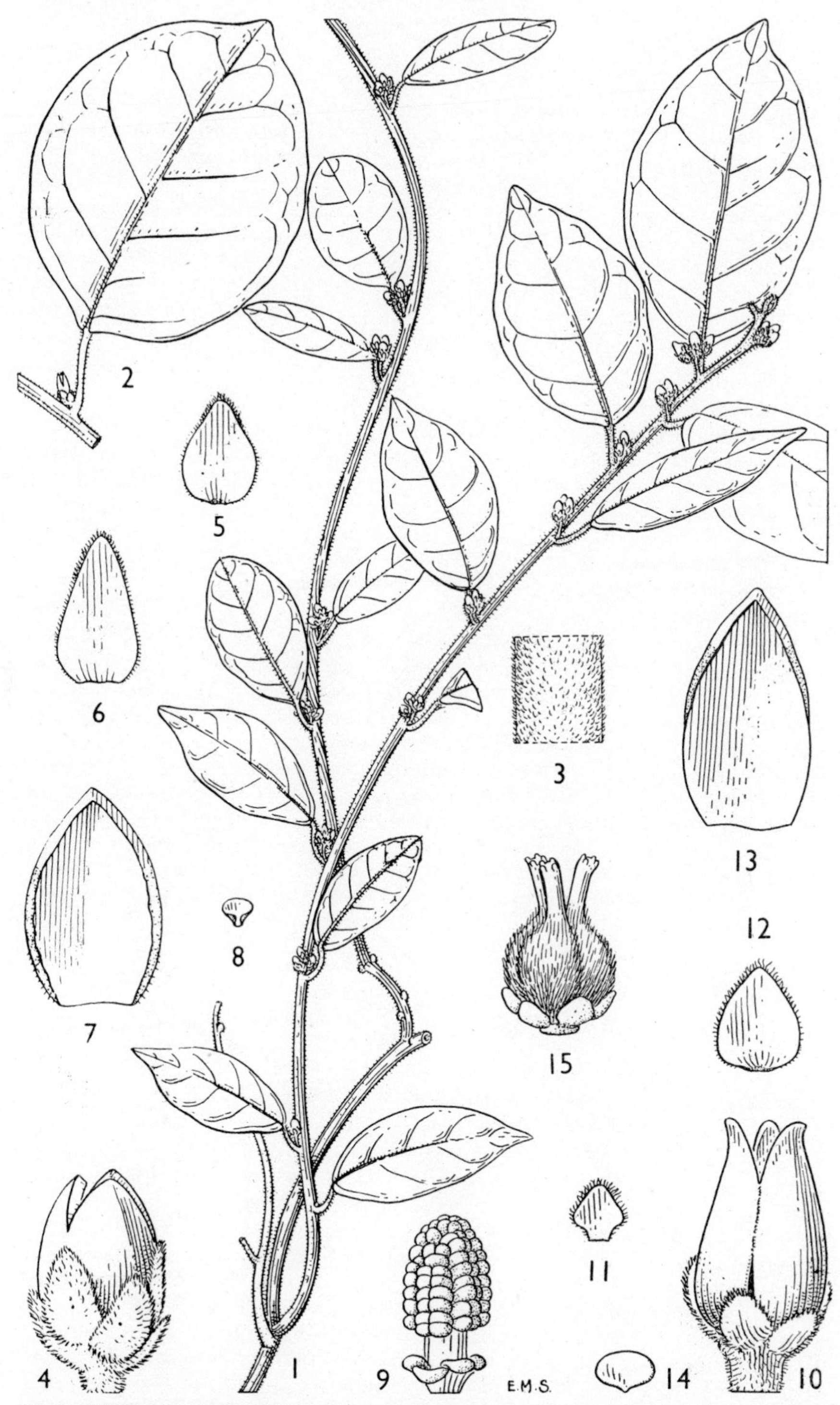

FIG. 1. *EPINETRUM EXELLIANUM*—**1,** habit, × 1 ; **2,** leaf with axillary inflorescence, × 1 ; **3,** hairs on stem, × 8 ; **4,** ♂ flower, × 8 ; **5,** outer sepal, × 12 ; **6,** median sepal, × 12 ; **7,** inner sepal, × 12 ; **8,** petal, × 12 ; **9,** stamens and petals, × 12 ; **10,** ♀ flower, × 8 ; **11,** outer sepal, × 12 ; **12,** median sepal, × 12 ; **13,** inner sepal, × 12 ; **14,** petal, × 12 ; **15,** carpels and petals, × 12. 1, 4–9, from *Chandler* 2171 ; 2–3, 10–15, from *Brown* 350.

1. EPINETRUM

Hiern in Cat. Afr. Pl. Welw. 1 : 21 (1896)

Suffrutescent, twining liane. Leaves simple, petiolate. Male inflorescences of small 1–4-flowered condensed cymules, solitary or two together, subsessile or pedunculate, axillary. Male flowers with 9–12 sepals, the 6–9 outer ones imbricate, bract-like, the 3 inner ones much larger than the outer ones, coriaceous, valvate, united to form a false corolla ; petals 6 or less, truncate-reniform, fleshy ; synandrium 15–30-locular, conical, stipitate ; anther-thecae with transverse dehiscence. Female flowers axillary, usually solitary ; sepals and petals more or less similar to those of the ♂ flowers : carpels 4–12, hairy. Drupes ovoid-ellipsoid ; exocarp echinulate, endocarp coriaceous and smooth. Seeds without endosperm.

Leaves completely glabrous, elliptic or slightly obovate, rounded at base, rounded and abruptly apiculate at apex, apiculus up to 1 cm. long . . . 1. *E. apiculatum*
Leaves sparsely puberulous except for the puberulous to tomentellous nerves, rarely glabrescent, rounded at base, ± rounded to obtuse at apex . . . 2. *E. exellianum*

1. **E. apiculatum** *Troupin* in B.J.B.B. 25 : 131 (1955). Type : Tanganyika, E. Usambara Mts., Bulwa, *Soleman in A.H.* 6006 (K, holo. !, EA, iso. !)

Liane with brownish-black glabrous branchlets. Leaves with slender glabrous petioles 1–1·5 cm. long ; blade elliptic or slightly obovate, rounded at base, rounded and abruptly apiculate at apex with an apiculus up to 1 cm. long, 4–9 cm. long, 3–5 cm. wide, glabrous and dark brown on both sides ; nerves in 4–6 pairs. Male inflorescences of 2–4-flowered condensed cymules, apparently solitary ; peduncle very short ; pedicels absent. Male flowers with 12 sepals, the 9 outer ones obovate-triangular to elliptic, the smallest ones 0·5 mm. long and wide, the biggest ones 2 mm. long and 1·2 mm. wide, pubescent, the three inner ones ovate-elliptic, 5–6 mm. long and 2–3 mm. wide, subcoalescent, glabrescent, blackish ; petals 6, 0·5 mm. long or less ; synandrium 30-locular, 3–4 mm. long, on a stipe 1–1·5 mm. long. Female inflorescences, ♀ flowers and fruits unknown.

TANGANYIKA. E. Usambara Mts., Bulwa, Nov. 1921 (♂ fl.), *Soleman in A.H.* 6006 !
DISTR. **T3** ; known only from Bulwa
HAB. Lowland rain-forest, about 1000 m.

SYN. [*E. undulatum* sensu Brenan & Greenway, T.T.C.L. : 327 (1949), *pro parte, quoad spec. Soleman in A.H.* 6006, *non* Hiern]

2. **E. exellianum** *Troupin* in B.J.B.B. 25 : 132 (1955). Type : Ruanda-Urundi, Mosso, Kininya, *Michel* 3154 (BR, holo. !)

Liane with young branchlets tomentellous, tawny-yellow ; mature branchlets sparsely puberulous to glabrescent. Leaves with yellowish-tomentellous petiole 0·8–1·5 cm. long ; blade elliptic, rounded at base, more or less rounded to obtuse at apex, 3·5–7 cm. long, 2·5–4·5 cm. wide, sparsely puberulous except for the densely puberulous to tomentellous nerves, sometimes glabrescent on both sides. Male inflorescences of 2–4-flowered condensed cymules, solitary or paired ; peduncles 0·5–3 mm. long or less ; pedicels absent. Male flowers with 9 sepals, the 3 outer ones ovate-triangular, 1·5 mm. long and 1 mm. wide, the 3 median suboblong, 2–2·5 mm. long and 1 mm. wide, all rusty-puberulous, the 3 inner ovate-elliptic, 4·5 mm. long and 1·5–2 mm. wide, glabrescent ; petals 6, 0·7 mm. long, 0·7–1 mm. wide ; synandrium 25–30-locular, 3·5–4 mm. long, on a stipe 1·5 mm. long. Female flowers solitary ; pedicels 0·5–2·5 mm. long ; carpels 3–6, 1·5–2·5 mm. long, densely puberulous, yellowish. Drupes unknown. Fig. 1, p. 3.

UGANDA. Kampala, Entebbe road, Feb. 1938 (♂ fl.), *Chandler* 2171 ! ; Entebbe, Dec. 1945 (young fr.), *Hansford* 3710 !
DISTR. U4 ; Ruanda-Urundi
HAB. Lowland rain-forest, about 1250 m.

SYN. [*E. undulatum* sensu Diels in E.P. IV. 94 : 95 (1910), & Brenan & Greenway, T.T.C.L. : 327 (1949), & Troupin in F.C.B. 2 : 221 (1951), *quoad Mildbraed* 565, *non* Hiern]

2. TRICLISIA

Benth. in G.P. 1 : 39 (1862)

Pycnostylis Pierre in Bull. Soc. Linn., Paris, n.s. : 81 (1898)

Robust, twining lianes. Leaves simple, petiolate ; blade entire, rarely serrate. Male inflorescences of corymb-like cymules or of small, many-flowered panicles. Male flowers with 9–18 sepals, all densely pubescent outside, the outer ones bract-like, the inner ones suborbicular and somewhat united ; petals 6, very small or absent, glabrous and somewhat fleshy ; stamens 3–6, free ; anther-thecae with obliquely longitudinal dehiscence ; a tuft of rusty hairs instead of the gynoecium. Female inflorescences similar to the ♂. Female flowers with sepals similar to those of the ♂ ; petals 3–6 or absent ; staminodes reduced ; carpels 6–40, pubescent, narrowed into a cylindrical style. Drupes obovoid, flattened, stipitate ; exocarp usually velvety, sparsely puberulous ; endocarp rugose and fibrous-hairy. Seeds without endosperm.

Leaf-margins always entire, sometimes slightly emarginate at apex :
 Petiole 2–10 cm. long ; leaves 7–18 cm. long, 5–13 cm. wide, dark brown when dried, rigid to subcoriaceous ; nerves rusty-tomentellous to puberulous beneath . . . . 1. *T. sacleuxii*
 Petiole 2–4 cm. long ; leaves 4·5–7·5 cm. long, 4–8 cm. wide, whitish-green when dried, papery ; nerves sparsely puberulous beneath . . . . . 2. *T. sp. A*
Leaf margins sometimes furnished with apiculi formed by the excurrent lateral nerves, membranous, dark brown when dried . . . . 3. *T. sp. B*

1. **T. sacleuxii** (*Pierre*) *Diels* in E.P. IV. 94 : 72 (1910). Type : Zanzibar Is., Mhamba, *Sacleux* 1874 (P, syn. !)

Liane with young branchlets rusty-tomentellous. Leaves tomentose to glabrescent, with petioles 2–10 cm. long ; blade elliptic, ovate-elliptic or ovate-lanceolate, rounded to subcordate at base, obtuse to shortly cuspidate or longly acuminate at apex, rigid to subcoriaceous, 7–18 cm. long, 5–13 cm. wide, glabrescent above, rusty-tomentose to puberulous on the nerves beneath ; nerves in 3–6 pairs. Male inflorescences axillary or springing from the old stems, of pedunculate cymules arranged in corymb-like clusters ; peduncles 0·5–1 cm. long, greyish-tomentellous. Male flowers with greyish-tomentellous pedicels 1–2 mm. long furnished with 3 bracteoles ; sepals 12–18, the 9–12 outer 2–4 mm. long and 1·5–3·5 mm. wide, the 3–6 inner 5–7 mm. long and 3·4 mm. wide ; petals 0·5–1 mm. long ; stamens 6, 3·5 mm. long. Female inflorescences similar to the ♂ ; peduncles of the cymules up to 2 cm. long and becoming robust in fruit. Female flowers with 20–25 carpels about 2 mm. long. Drupes truncate at base, 1·7–3 cm. long, 1·5–1·7 cm. wide. Seeds 2–2·5 cm. long.

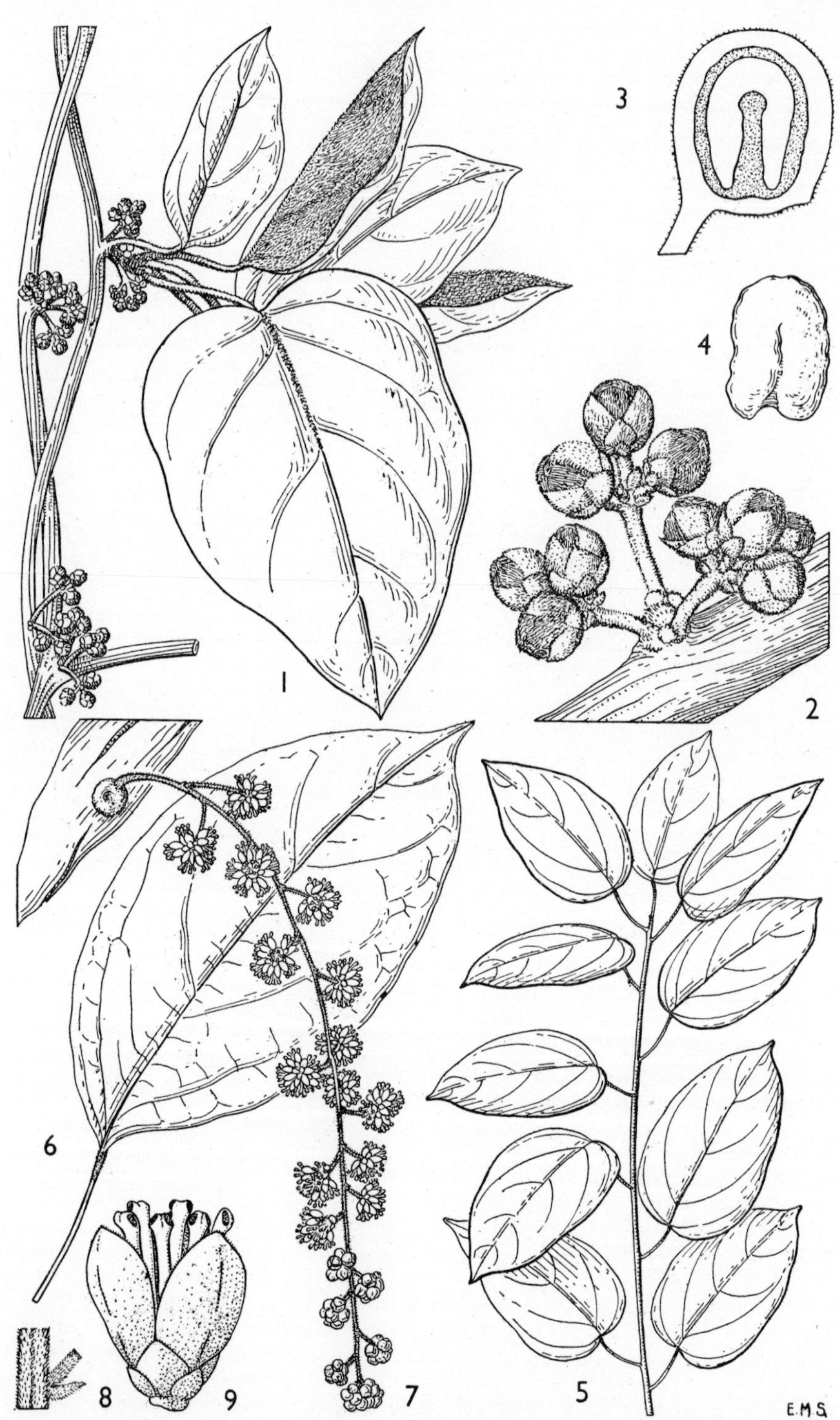

FIG. 2. *TRICLISIA SACLEUXII*—**1,** habit, × $\frac{2}{3}$; **2,** ♂ inflorescence, × 4 ; **3,** section of fruit, × $1\frac{1}{2}$; **4,** seed, × $1\frac{1}{2}$. *TILIACORA FUNIFERA*—**5,** leafy twig, × $\frac{1}{3}$; **6,** leaf, × $\frac{2}{3}$; **7,** ♂ inflorescence, × $\frac{2}{3}$; **8,** hairs on the inflorescence, × 4 ; **9,** ♂ flower, × 8. 1–2, from *Faulkner* 1084 ; 3–4, from *Drummond & Hemsley* 1446 ; 5, from *C. G. Rogers* 5322 ; 6, from *Drummond & Hemsley* 4774 ; 7–9, from *Harris* 150.

var. sacleuxii

Leaves ovate-elliptic to broadly elliptic, rounded to subcordate at base, obtuse to emarginate or shortly cuspidate at apex. Fig. 2/1–4.

KENYA. Teita Hills, Mboloeo Hill, Feb. 1953 (fr.), *Bally* 8575!
TANGANYIKA. Tanga District: Magunga Estate, Oct. 1952 (♂ fl.), *Faulkner* 1084! & Magunga, May 1916, *Braun in A.H.* 6420! & East Usambara Mts., Tongwe, Mlinga Hill, Mar. 1953 (fr.), *Drummond & Hemsley* 1446!
ZANZIBAR. Zanzibar Is., Ufufuma, Dec. 1930, *Vaughan* 1753!
DISTR. **K**7; **T**3, 6; **Z**; Middle Congo, Angola, Portuguese East Africa
HAB. Lowland rain-forest, riverine forest & moist shady places, 0–1200 m.

SYN. *Pycnostylis sacleuxii* Pierre in Bull. Soc. Linn., Paris, n.s.: 81 (1898)
Chondodendron ? *macrophyllum* Hiern in Cat. Afr. Pl. Welw. 1: 16 (1896). Type: Angola, Golungo Alto, *Welwitsch* 2307 (BM, holo.!, K, P, iso.!)
Welwitschiina macrophylla (Hiern) Engl. in E.J. 26: 416 (1899)
Triclisia welwitschii Diels in E.P. IV. 94: 69 (1910), *nom. nov.*, *non* Hiern (1896), *nom. illegit.* Type: as *Chondodendron* ? *macrophyllum* Hiern
T. lucida Exell & Mendonça in Consp. Fl. Angol. 1: 35 (1937), *nom. nov.* Type as *Chondodendron* ? *macrophyllum* Hiern
[*T. macrophylla* sensu Chev., F.V.A.O.F. 1: 113 (1938), *pro parte*, *quoad Chevalier* 4219, *non* Oliv.]

var. ovalifolia *Troupin* in B.J.B.B. 25: 134 (1955). Type: Tanganyika: Rungwe District: Mbaka Kilambo, *Stolz* 1604 (K, holo.!, BM, BR, IFI, P, iso.!)

Blade ovate-lanceolate, markedly cordate at base, longly acuminate at apex.

TANGANYIKA. Pangani District: Tongwe, Dec. 1928, *Murray* H24/28!; Rungwe District: Mbaka Kilambo, Oct. 1912, *Stolz* 1604! & Mar. 1913, *Stolz* 1922!
DISTR. **T**3, 7; Portuguese East Africa
HAB. Lowland rain-forest & riverine forest, 0–1700 m.

SYN. *Triclisia* sp., T.T.C.L. 328 (1949)—*Stolz* 1604 (K!)

2. T. sp. A

Liane with puberulous to glabrescent branchlets. Leaves with bright green petiole 2–4 cm. long; blade broadly elliptic to suborbicular, truncate or ± rounded at base, broadly obtuse to ± rounded at apex, 4·5–7·5 cm. long, 4–8 cm. wide, papery, bright green, glabrous on both sides except sometimes for the sparsely puberulous nerves beneath; nerves in 3–4 pairs. Inflorescences and fruits unknown.

TANGANYIKA. Pangani District: Bushiri Estate, *Faulkner* 769!
DISTR. **T**3; known only from this gathering
HAB. Lowland rain-forest, 80 m.

3. T. sp. B

Liane with puberulous to rusty-tomentellous branchlets. Leaves with ± densely puberulous to rusty-tomentellous petioles 5–9 cm. long; blade broadly ovate, either entire or with one or two lateral apiculi, rounded at base, caudate and apiculate at apex, 8–15 cm. long, 5–12 cm. wide, very sparsely puberulous on both sides except for the ± densely puberulous nerves. Male and ♀ inflorescences unknown. Drupes flattened, ± rounded, truncate at base, 1–1·5 cm. long, 0·6–0·8 cm. thick. Stipe about 0·5 cm. long; exocarp rusty-puberulous.

UGANDA. Toro District: S. Kibale Forest, Dec. 1948, *Loveridge* 244!
DISTR. **U**2; known only from this gathering
HAB. Lowland rain-forest, 1500 m.

3. TILIACORA

Colebr. in Trans. Linn. Soc. 13: 53 (1822)

Glossopholis Pierre in Bull. Soc. Linn., Paris, n.s.: 82–85 (1898)

Robust lianes. Leaves simple, petiolate; blade entire, penninerved. Male inflorescences of false racemes of condensed cymules, few-flowered, axillary or from the old stems, or of axillary, solitary, sometimes 1-flowered cymules.

Male flowers with 6–12 sepals, glabrous or rarely partly puberulous, the outer bract-like, the inner much larger, obovate or elliptic, somewhat fleshy or coriaceous ; petals 3–6 ; stamens 3–9, free or longly united ; anthers introrse ; thecae dehiscing longitudinally. Female inflorescences similar to the ♂, or more simple, sometimes as spikes of solitary flowers. Female flowers with sepals and petals similar to those of the ♂ ; staminodes absent ; carpels 6–30, borne on an apparent gynophore. Drupes ovate, stipitate ; the remains of the stigma visible near the stipe ; exocarp glabrescent, smooth or verrucose, endocarp compressed, woody, furrowed. Endosperm of seeds ruminate, sparse or absent.

Female inflorescences of spikes of solitary flowers ; leaves longly acuminate, apiculate at apex, apiculus up to 2 cm. long . . . . 1. *T. kenyensis*
Female inflorescences of false racemes of 3–9-flowered cymules :
  Carpels 15–30 ; leaves broadly ovate-elliptic ; ♂ flowers with included stamens . . . . . 2. *T. latifolia*
  Carpels 8–15 ; leaves lanceolate to ovate-elliptic ; ♂ flowers with exserted stamens :
    Leaves ovate-elliptic to broadly elliptic, subcordate or rounded to somewhat obtuse at base, 5–20 cm. long and 3–10 cm. wide, puberulous to glabrescent 3. *T. funifera*
    Leaves usually ovate-lanceolate, obtuse to cuneate at base, not more than 8 cm. long and 5 cm. wide, often narrower, completely glabrous . . . 4. *T. sp.*

1. **T. kenyensis** *Troupin* in B.J.B.B. 25 : 133 (1955). Type : Kenya : N. Kavirondo District : Kakamega Forest, *Drummond & Hemsley* 4783 (K, holo. !, BR, iso. !)

Twining liane with glabrous branchlets. Leaves : petiole 1·5–2 cm. long, filiform, smaller in adult leaves ; blade ovate-oblong, slightly cuneate at base, longly acuminate-apiculate at summit (apiculus up to 2 cm. long), slightly undulate on margins, 7·5–13 cm. long, 3–4 cm. wide, membranous, glabrous on both sides ; nerves 5–7 pairs. Male inflorescences and flowers unknown. Female inflorescences springing from the old stems, of spikes up to 15 cm. long of solitary flowers. Female flowers with 12 sepals, the 9 outer ± broadly triangular, 1·2–3 mm. long, 1–2·5 mm. wide, sparsely ciliolate, the 3 inner suborbicular, 2·5–3 mm. long and wide ; petals 6, ± oblong, ± acute to subapiculate at apex, about 1 mm. long, thickened on margins, glabrous ; carpels 10–15, 1–1·5 mm. long. Fruits unknown.

KENYA. N. Kavirondo District : Kakamega Forest, near the forester's house, just off Kakamega–Kaimosi road, Oct. 1953 (♀ fl.), *Drummond & Hemsley* 4783 !
DISTR. **K**5 ; known only from Kakamega Forest
HAB. Upland rain-forest, 1600 m.

2. **T. latifolia** *Troupin* in B.J.B.B. 25 : 133 (1955). Type : Uganda, Ankole District : Kalinzu Forest, *Eggeling* 3670 (K, holo. !, BR, iso. !)

Liane with sparsely puberulous to glabrescent branchlets. Leaves with slender petiole 1·5–3 cm. long, and blackish at apex ; blade broadly ovate-elliptic, rounded to subtruncate at base, acuminate to subapiculate at apex, 8–14 cm. long, 5–8 cm. wide, papery ; nerves in 4–5 pairs, sparsely puberulous or glabrescent beneath. Female inflorescences of false racemes 8–10 cm. long clustered 2–5 together ; cymules 3–9-flowered, on peduncles 0·3–1·2 cm. long. Male flowers with 9–12 sepals, the 6–9 outer ones triangular, 1–3 mm. long and wide, ciliolate, the 3 inner ovate-oblong, 6–7 mm. long, 2–3 mm. wide ; petals 3, narrowed at base, 2·5–3 mm. long ; stamens 6, included, free or slightly united at base, 6–7 mm. long. Female inflorescences similar to the ♂. Female flowers with 9–12 sepals, the inner suborbicular, 4–5 mm. long and wide ; petals 6, ± clawed at base ; carpels 15–30, 1·5–2 mm. long ; gynophore 1·5–2 mm. long. Drupes 1–1·3 cm. long and 0·6–0·7 cm. wide ; stipe 0·7–1 cm. long.

Uganda. Bunyoro District : Budongo Forest, Sep. 1935 (♂ & ♀ fl.), *Eggeling* 2179 ! & 2225 !

Distr. **U**2 ; also in Portuguese East Africa

Hab. Lowland rain-forest, 1000–1500 m.

3. **T. funifera** (*Miers*) *Oliv.*, F.T.A. 1 : 44 (1868) ; Diels in E.P. IV. 94 : 64 (1910). Type : Nyasaland, Manganja Hills, *Meller* (K, holo. !)

Woody liane. Leaves : petiole slender, 1·5–5 cm. long, puberulous to glabrescent ; blade ovate-lanceolate, ovate-oblong or broadly ovate, subcordate to rounded or somewhat obtuse at base, obtuse to acute or acuminate at apex, 5–20 cm. long, 3–10 cm. wide, papery or coriaceous, glabrous ; nerves in 3–5 pairs, sometimes sparsely puberulous beneath. Male inflorescences either of axillary solitary cymules on peduncles 1–1·2 cm. long, or of 3–9-flowered cymules arranged in false racemes which are axillary or springing from the old stems, solitary or clustered, and up to 15 cm. long ; axes and peduncles puberulous. Male flowers with 6–9 sepals, the 3–6 outer ones triangular to orbicular, 0·8–1·5 mm. long and wide, thickened and ciliolate, the 3 inner obovate-elliptic, 3·5–4 mm. long, 1·8–2·3 mm. wide ; petals 6, clawed, thickened on margins, 1·5–2·5 mm. long ; stamens exserted, 3–5 mm. long, free or slightly united at base. Female inflorescences similar to the ♂. Female flowers with 6–9 sepals, the outer ones lanceolate to ovate, 1–2 mm. long, the inner suborbicular, 2·5 mm. long and wide ; petals (5–) 6, 1–1·5 mm. long ; carpels 8–12, about 1 mm. long. Drupes obovoid to nearly round, 5–7 mm. long ; stipe 1·5–3 mm. long. Fig. 2/5–9, p. 6.

Uganda. Kampala, Bombo road, Jan. 1938 (fl. buds), *Chandler* 2124 ! ; Entebbe, Kyewaga Forest, Sep. 1922 (♂ fl.), *Maitland* 160 !

Kenya. N. Kavirondo District : Kakamega Forest, Kakamega–Kaimosi road, Oct. 1953 (♂ fl.), *Drummond & Hemsley* 4774 ! & May 1933 (♂ fl.), *Dale* 3130 !

Tanganyika. Mpwapwa, Dec. 1934 (fr.), *H. E. Hornby* 624 ! ; Morogoro District : Nguru Mts., Manyungu Forest, Mar. 1953 (fr.), *Drummond & Hemsley* 1836 ! ; Lindi, Oct. 1934, *Schlieben* 5514 !

Distr. **U**4 ; **K**4–5 ; **T**2, 4–8 ; Togo, Gold Coast, Nigeria, Belgian Congo, Angola, Northern and Southern Rhodesia, Nyasaland, Portuguese East Africa

Hab. Lowland & upland rain-forest, riverine forest and moist shady places in woodland, 220–1250 m.

Syn. *Hypserpa funifera* Miers in Contr. Bot. 3 : 104 (1871)
Tiliacora warneckei [Engl. ex] Diels in E.P. IV. 94 : 64 (1910). Type : Togo, Lome, *Warnecke* 221 (B, holo. !, BM, BR, EA, K, P, iso. !)
T. pynaertii De Wild., Bull. Inst. Roy. Colon. Belge 2 : 573 (1931). Type : Belgian Congo, Eala, *Pynaert* 615 (BR, holo. !)
T. glycosmantha Diels in N.B.G.B. 11 : 662 (1932). Type : Tanganyika, Njombe District, Lupembe, *Schlieben* 1365 (B, holo. †, BR, lecto. !, BM, isolecto. !)
T. johannis Exell, in J.B. 73, Suppl. Addend. : 7 (1935). Type : Angola, Cabinda, *Gossweiler* 9046 (BM, holo. ! ; K !, B †, LISJC, iso.)
Triclisia sp., T.T.C.L. 328 (1949)—*Hornby* 624 (K !)

The variation within this species is shown particularly in the shape and consistency of the leaves ; this has led to various alleged taxa being distinguished. Belgian Congo specimens generally have elongate and very coriaceous leaves.

4. **T. sp.**

Liane with glabrous stems and branchlets. Leaves : petiole slender, 1·5–3 cm. long, blackish, sparsely puberulous to glabrescent ; blade narrowly ovate-lanceolate to lanceolate, obtusely cuneate at base, acuminate at apex, 5–7 (–8) cm. long, 1·8–3·5 (–5) cm. wide, glabrous on both sides, slightly discolorous ; nerves in 4–6 pairs. Male inflorescences and ♂ flowers unknown. Female inflorescences axillary (according to the collector), in false racemes 10–15 cm. long on 2–4 mm.-long peduncles, composed of 3–9-flowered cymules ; axes and peduncles puberulous. Female flowers with 9 sepals, the 6 outer triangular, 1–1·5 mm. long and wide, the 3 inner ovate, 5–6 mm. long and 2–2·5 mm. wide, glabrous, somewhat fleshy ; petals 6, 2–2·5 mm. long ; carpels 10–15, about 1 mm. long ; gynophore 3–5 mm. long. Infructescences with peduncles of the cymes up to 8 mm. long and the gynophores up to 6 mm. long. Fruit unknown.

KENYA. Masai District : Emali Hill, Aug. 1940 (♀ fl.), *van Someren* 91 !
DISTR. **K**6 ; not known elsewhere
HAB. Upland dry evergreen forest, about 1900 m.

4. COCCULUS

DC., Syst. 1 : 515 (1817), *nom. conserv.*

Menispermum L., Gen. Pl., ed. 5 : 518 (1754), *pro parte, non typica*
Cebatha Forsk., Fl. Aegypt.-Arab.: 171 (1775)
Leaeba Forsk., Fl. Aegypt.-Arab.: 172 (1775)
Holopeira Miers in Ann. Mag. Nat. Hist., ser. 2, 7 : 52 (1851)

Prostrate or erect lianes or suffrutices ; branchlets rambling, or reduced to cladodes[1]. Leaves simple, various in form, entire or lobed. Male inflorescences of cymules which are either axillary and clustered 1–3 together or solitary and arising from leafless branches, or grouped in ± condensed clusters on the cladodes[1] ; pedicels present or absent. Male flowers with 6 sepals, the 3 outer ones reduced, the 3 inner concave ; petals 6, furnished at base with ± fleshy inflexed auricles surrounding the stamen-filaments, often bifid or deeply emarginate at apex ; stamens 6–9, free ; anther-thecae with transverse dehiscence. Female inflorescences similar to the ♂ or more simple or reduced to solitary or clustered flowers. Female flowers with sepals similar to those of the ♂ ; base of petals much less inflexed ; staminodes 6, linear-filiform or absent ; carpels 3–6, subovoid and compressed, with cylindrical style and recurved-spathulate stigma. Drupes obovoid or flattened and rounded ; the remains of the style visible near the base. Seeds with sparse endosperm.

Leaf-blade usually glabrous or very slightly puberulous ; basal nerves 3 ; petiole 0·2–1 cm. long ; fruit with septum of the condyle[2] not perforated . . . . 1. *C. pendulus*
Leaf-blade ± densely tomentellous to puberulous ; basal nerves 5 ; with petiole 0·5–2·5 cm. long ; fruit with septum of the condyle perforated . . . . 2. *C. hirsutus*

1. **C. pendulus** (*J. R. & G. Forst.*) *Diels* in E.P., IV. 94 : 237, fig. 78 (1910) ; F.W.T.A. 1 : 79 (1927) & ed. 2, 1 : 76 (1954) ; Chev., F.V.A.O.F. 1 : 98 (1938) ; Cufodontis in B.J.B.B. 24, suppl. : 113 (1954). Type : Cape Verde Is., S. Tiago, *Forster* (BM, holo. !)

Liane with stem up to 15 cm. in diameter at ground-level ; young branchlets slender, puberulous. Leaves with petiole 0·2–1 cm. long ; blade oblong-lanceolate, sometimes ovate in the lower leaves, base broadened, cuneate, ± rounded or trilobed-hastate, apex obtuse and mucronulate or sometimes emarginate, 1·3–5 cm. long, 0·5–1·8 cm. wide, usually glabrous

[1] *C. balfourii* Schweinf., confined to the island of Socotra.
[2] The condyle is a prominent enlargement of the placenta and forms a hollow chamber within the cavity of the cell round which the seed is moulded.

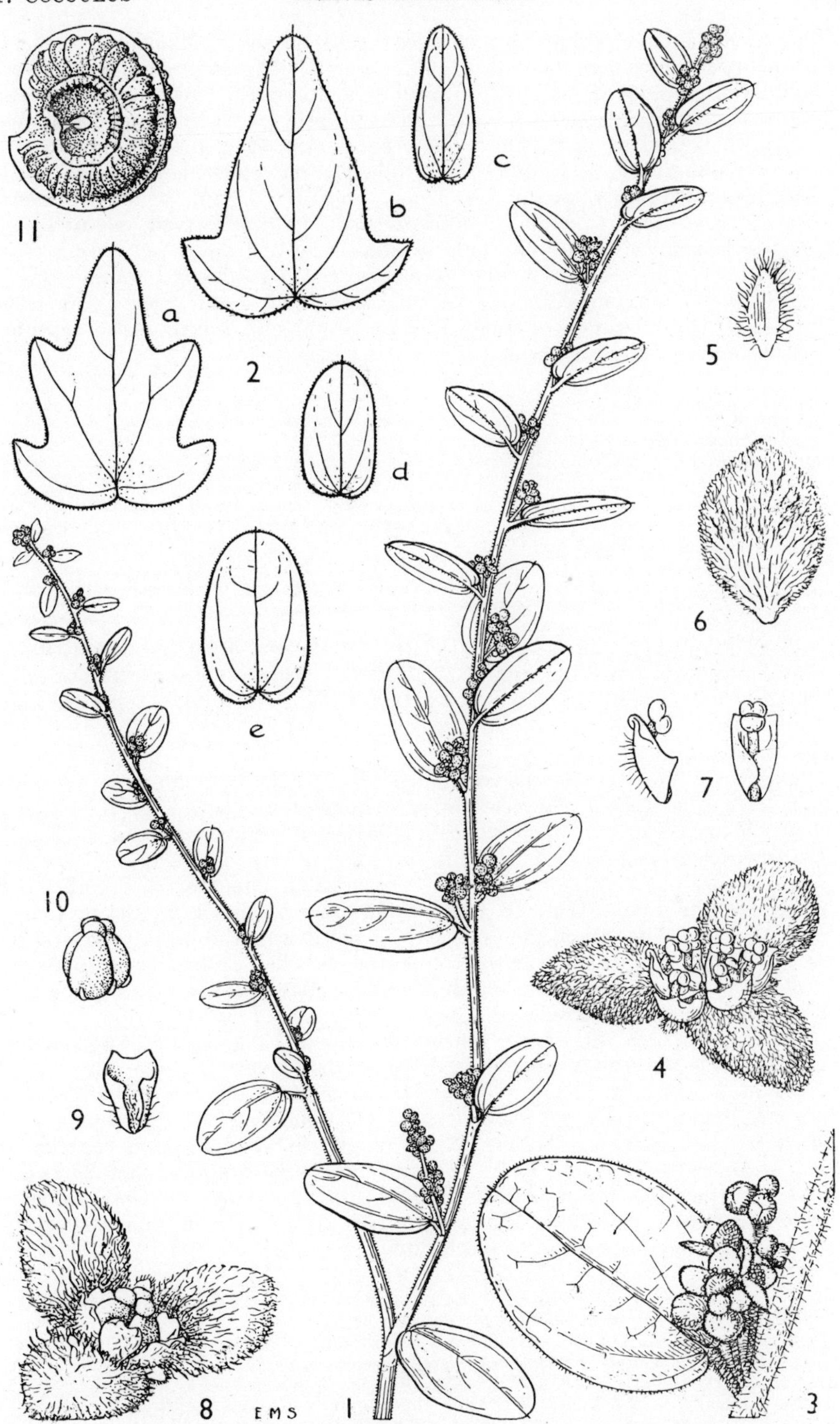

Fig. 3. *COCCULUS HIRSUTUS*—**1,** habit, × 1 ; **2** a–e, leaf variation, × 1 ; **3,** ♂ inflorescence, × 3 ; **4,** ♂ flower, × 10 ; **5,** outer sepal, × 10 ; **6,** inner sepal, × 10 ; **7,** petals and stamens, × 10 ; **8,** ♀ flower, × 8 ; **9,** petal, × 10 ; **10,** carpels and staminodes, × 10 ; **11,** seed, × 5. 1, 2d, 3–10, from *Robinson* 313 ; 2a–c, from *L. Scott* ; 2e, from *Greenway* 4483 ; 11, from *Trapnell* 1124.

or very slightly puberulous on both sides ; basal nerves 3. Male inflorescences of many-flowered, axillary, fasciculate or ± congested cymules, rarely solitary, (0·5–1·5) –2 cm. long ; peduncles up to 1·3 cm. long ; pedicels short or absent. Male flowers with ovate-elliptic, ± fleshy to membranous sepals thickened at base, the 3 outer 1–1·5 mm. long, 0·4–0·7 mm. wide, sparsely puberulous and ciliate, the 3 inner larger, puberulous to glabrous ; petals subovate to obovate, 0·7–2 mm. long, 0·5–1 mm. wide ; stamens 0·8–1·5 mm. long. Female inflorescences of few-flowered, solitary or clustered cymules 0·7–1·3 cm. long ; pedicels up to 1 cm. long ; bracteoles 1 or 2, 0·5 mm. long. Female flowers with staminodes 0·5–1 mm. long ; carpels 0·8–1 mm. long. Drupes 4–7 mm. long and 4–5 mm. wide ; endocarp ribbed on the lateral faces and without a prominent dorsal crest ; septum of the condyle[1] not perforated.

KENYA. Turkana District : Kateruk River, July 1954, *Hemming* 348 !
DISTR. **K**2 ; widely spread from the Cape Verde Is. & Spanish Sahara to Somalia ; also in Socotra, Algeria, Egypt & Arabia
HAB. Semi-desert scrub and deciduous bushland, about 700 m.

SYN. *Epibaterium pendulum* J. R. & G. Forst., Char. Gen. : 108, t. 54 (1776)
Menispermum leaeba Del., Fl. Egypt. : 140, t. 51, fig. 2–3 (1813). Type : Egypt, near Cairo, *Delile* (MPU, syn.)
Cocculus leaeba (Del.) DC., Syst. : 529 (1817) ; F.T.A. 1 : 44 (1868)
Cebatha pendula (J. R. & G. Forst.) O. Ktze., Rev. Gen. 1 : 9 (1891)

2. **C. hirsutus** (*L.*) *Diels* in E.P. IV. 94 : 236 (1910) ; T.T.C.L.: 326 (1949) ; Cufodontis in B.J.B.B. 23, suppl : 112 (1953). Type : "Indes Orientales," Plukenet's drawing [Amalth. Bot. 61, t. 384, fig. 7 : (1705)] ; specimen in *Herb. Sloane* (BM !)

Liane reaching several metres in length ; young branchlets ± densely pubescent-tomentose. Leaves yellowish-tomentose, petiole 0·5–2·5 cm. long ; blade various in shape, that of the leaves in the lower part of the main branches clearly 3–5-lobed, that of the other leaves narrowly to broadly ovate, ovate-oblong or obovate, base broadened, cuneate, ± rounded or rarely cordate, apex obtuse to rounded and mucronulate, 4–8 (–9) cm. long, (2·5–6) –7 cm. wide, bract-like and then 0·5 cm. long at the ends of the lateral and flowering branchlets, densely tomentellous when young, later sparsely tomentellous to glabrescent ; basal nerves 5. Male inflorescences of many-flowered cymules clustered 2–3 together or rarely solitary, 1–2·5 cm. long ; peduncle up to 1·5 cm. long ; pedicels 0·5–1 mm. long. Male flowers with long-hairy sepals, the 3 inner broadly ovate or obovate, 1·5–2·5 mm. long, 1·7–2 mm. wide, the 3 outer oblong to lanceolate, 1·4–2 mm. long, 0·4–0·8 mm. wide ; petals ovate-oblong, 0·5–1·5 mm. long, 0·3–0·6 mm. wide, sparsely pubescent to glabrescent ; stamens 0·7–1 mm. long. Female inflorescences 0·5–2·5 cm. long. Female flowers with staminodes 0·5 cm. long ; carpels 0·7–1 mm. long. Drupes 4–8 mm. long, 4–5 mm. wide ; endocarp clearly ribbed on the lateral faces and with a prominent dorsal crest ; septum of the condyle[1] perforated. Fig. 3, p. 11.

KENYA. Northern Frontier Province : Sololo, Aug. 1952 (♂ fl.), *Gillett* 13681 ! Machakos District : Kibwezi, July 1909 (♂ fl.), *Scheffler* 347 !
TANGANYIKA. Moshi District : Himo River, Jan. 1936 (♀ fl.) *Greenway* 4483 ! ; Chunya District : Galula, Feb. 1934 (♂ fl.), *Michelmore* 984 !
ZANZIBAR. Without locality, probably Zanzibar Is., *Vaughan* 1178 !
DISTR. **K**1, 4, 6 ; **T**2, 4–6 ; **Z** ; from the Sudan and Eritrea south to Natal and South West Africa ; also in Asia from central Arabia to southern China
HAB. Bushland & semi-desert scrub, up to 1140 m.

[1] See footnote (2) on page 10.

Syn. *Menispermum hirsutum* L., Sp. Pl. : 341 (1753)
M. villosum Lam., Encycl. Méth. Bot. 4 : 97 (1797), *non* Roxb. (1832). Type : "Indes Orientales," *Sonnerat* (P–LA, holo.)
Cocculus villosus (Lam.) DC., Syst. 1 : 525 (1817) ; F.T.A. 1 : 45 (1868)
Cebatha hirsuta (L.) O. Ktze., Rev. Gen. 1 : 9 (1891) ; Dur. & Schinz, Consp. Fl. Afr. 1 (2) : 47 (1898)

5. DIOSCOREOPHYLLUM

Engl. in P.O.A. C : 181 (1895)

Rhopalandria Stapf in K.B. 1898 : 71 (1898)
Dioscoreopsis O. Ktze. in Post & O. Ktze., Lex. Phan.: 176 (1904)

Herbaceous, twining lianes. Leaves simple, longly petiolate, entire or lobed, often of various shapes. Male inflorescences of axillary, longly pedunculate racemes. Male flowers with 6–8 sepals in 2 whorls ; petals absent ; stamens 3–6, fused into a cylindrical or subglobular, sometimes flattened, sessile or stipitate synandrium ; anther-thecae oblong, parallel, dehiscing longitudinally. Female inflorescences similar to the ♂. Female flowers with 6 sepals ; petals absent ; carpels 3–6, with thickened, recurved stigmas. Drupes subovoid, topped by the remains of the style and stigma ; exocarp smooth and sometimes shining ; endocarp crustaceous, verrucose, or covered on upper side with short prickles much enlarged at base. Seeds with thick fleshy endosperm.

Ten species have been recognized in this genus, based on characters of leaf-shape and indumentum, which, however, have not proved sufficiently distinct to justify specific separation ; the variation shown by the leaves is particularly noteworthy. On the other hand, the shape and structure of the androecium seem to give good distinguishing characters, while the various types of indumentum are of not more than varietal significance. On this basis we are able to recognize in the whole of Africa only three species, of which two have infraspecific taxa.

D. volkensii *Engl.* in P.O.A. C : 182 (1895). Type : Tanganyika, E. Usambara Mts., Nderema, *Volkens* 109 (B, holo.!)

Liane with ± slender, sparsely pubescent or hirsute branchlets. Leaves with petiole 8–15 cm. long ; blade varying from entire, ovate-triangular, sagittate-cordate at base, sharply acuminate at apex to toothed or to 3–5-lobed with lobes ± irregular, sharply acuminate or apiculate at apex, 9–18 cm. long, 5·5–15 cm. wide, herbaceous or submembranous, sparsely hairy to glabrescent ; basal nerves 5–7, palmate. Male inflorescences up to 15 cm. long, on a peduncle 6 cm. long or more ; axis hairy or pubescent ; bracts 1·5–3 mm. long ; pedicels 2–4 mm. long. Male flowers with oblong-obovate sepals, 2–4 mm. long, 1–2 mm. wide, sparsely pubescent or hairy ; synandrium subglobular or slightly flattened, 1–2 mm. long, on a slender stipe up to 2 mm. long. Female inflorescences 10–18 cm. long. Female flowers with carpels 1·5–2 mm. long. Drupes 0·6–1·2 cm. long ; peduncle glabrescent, about 1 cm. long. Seeds 1–2·5 cm. long.

var. **volkensii**

Plant sparsely clothed with slender, pale, non-rusty hairs. Fig. 4, p. 14.

Kenya. Kwale District : Shimba Hills, Magadara, Aug. 1953 (fr.), *Drummond & Hemsley* 4037 !
Tanganyika. E. Usambara Mts., Amani, *Soleman in A.H.* 5999 ! & Lewa, July 1941 (♂ fl.), *Yusufu bin Mohamedi in A.H.* 9097 ! ; Ulanga District : Mahenge, Feb. 1932 (♂ fl.), *Schlieben* 1749 !
Zanzibar. Pemba Is., *Vaughan* 604 !
Distr. **K**7 ; **T**3, 6, 7 ; **P** ; also in West Africa from Sierra Leone to Nigeria
Hab. Lowland rain-forest, up to 1140 m.

Syn. *D. volkensii* Engl. in P.O.A. C : 182 (1895) ; Diels in E.P. IV. 94 : 183 (1910), *pro parte, excl. Swynnerton* 100 & 100a

FIG. 4. *DIOSCOREOPHYLLUM VOLKENSII*, from *Faulkner* 577—**1,** leaf with ♂ inflorescence, × 1; **2,** ♂ flower, × 7; **3,** leaf with ♀ inflorescence, × 1; **4,** ♀ flower, × 7; **5,** infructescence, × 1; **6,** fruit, × 2; **7,** section of the drupe showing the endocarp, × 3.

D. tenerum Engl. in E.J. 26 : 407 (1899) ; Diels in E.P. IV. 94 : 181 ; F.W.T.A. 1 : 71 (1927) ; Chev., F.V.A.O.F. 1 : 108 (1938). Type : Sierra Leone, *Afzelius* (B, holo. !, K, iso. !)
D. tenerum Engl. var. *tenerum*, F.W.T.A., ed. 2, 1 : 73 (1954)

The var. *fernandense* (Hutch. & Dalz.) Troupin is characterized by the presence of stiff, rusty-brown hairs, and is recorded from southern Nigeria and Fernando Po.

6. JATEORHIZA

Miers in Hook., Niger Fl. 212 (1849)

Somewhat woody liane with strong indumentum. Leaves longly petiolate, 3–5 (–7)-lobed. Male inflorescences of elongate axillary panicles, the lateral axes bearing 3–7-flowered clusters. Male flowers with 6 sepals, the 3 outer elongate to elliptic, the 3 inner obovate ; petals 6, somewhat concave, mostly abruptly bent inwards at apex and with their margins recurved inwardly and enveloping the androecium ; stamens 6, free or connate ; anthers introrse, globular ; thecae with transverse dehiscence. Female inflorescences of axillary racemes. Female flowers with sepals and petioles ± similar to those of the ♂ ; staminodes 6, tongue-shaped. Carpels 3, subovoid ; styles small, recurved, the broad stigma produced into 2–3-cleft lamellae. Drupes ovoid or subovoid ; exocarp strigose-hispid or setulose ; endocarp ovoid, flattened, the ventral side ± smooth, the dorsal side clothed with numerous fibrillose hairs. Seeds with fleshy, ruminate endosperm.

The original spelling of the name of this genus was wrongly altered by Engler (1899) to *Jatrorrhiza*. Various authors (Diels ; Th. Durand & Schinz ; Hutchinson & Dalziel) have adopted this change which, however, is not justified by the Nomenclatural Code.

J. palmata (*Lam.*) *Miers* in Hook., Niger Fl. 214 (1849) ; Dur. & Schinz, Consp. Fl. Afr. 1 (2) : 46 (1898) ; Diels in E.P. IV. 94 : 166 (1910) ; T.T.C.L. : 327 (1949). Type : cultivated in Mauritius, *Commerson* (P–LA, holo. !, P, iso. !)

Liane with branchlets densely pubescent at first, later strigose. Leaves with strigose petioles 18–25 cm. long ; blade broadly rounded, deeply cordate at base, generally with 5 broadly ovate lobes, acuminate at apex, sometimes angular, 15–35 cm. long, 16–40 cm. wide, membranous, clothed with strigose hairs on both sides, rarely glabrescent ; basal nerves 5–7, palmate. Male inflorescences 40 cm. long ; lateral branches 2–10 cm. long ; main axis strigose ; secondary axes sometimes glabrous, with a linear-lanceolate ciliate bract at base ; pedicels absent. Male flowers with greenish sepals, 2·7–3·2 mm. long and 1·3–1·6 mm. wide ; petals 1·8–2·2 mm. long ; stamens free, slightly adnate to the base of the petals, 1–1·8 mm. long. Female inflorescences 8–10 cm. long. Female flowers with carpels 1–1·5 mm. long, rusty-pubescent. Drupes 2–2·5 cm. long, 1·5–2 cm. wide.

KENYA. Tana River District : Mambosasa, Feb. 1929 (fl. buds), *Graham* 1797 !
TANGANYIKA. Handeni District : without locality, *Omari bin Chambo* 1 ! ; Lindi District : Lake Lutamba, Jan. 1935 (♂ fl.), *Schlieben* 5851 !
DISTR. **K**7 ; **T**3, 7, 8 ; Gold Coast, Nyasaland, Portuguese East Africa, Mauritius
HAB. Lowland rain-forest and riverine forest, 0–1500 m.

SYN. *Menispermum palmatum* Lam., Encycl. Méth. 4 : 99 (1797)
Cocculus palmatus (Lam.) DC., Syst. 1 : 522 (1817)
Menispermum columba Roxb., Fl. Ind. 3 : 807 (1832). Type : cultivated at Calcutta, originally from Portuguese East Africa, *Roxburgh* (BR, lecto. !)
Jateorhiza columba (Roxb.) Oliv., F.T.A. 1 : 42 (1868)
J. miersii Oliv., F.T.A. 1 : 42 (1868). Type : cultivated in Mauritius, *Bojer* (K, holo. !, W, iso.)
Chasmanthera columba (Roxb.) Baill., Hist. Pl. 3 : 30, fig. 16, 17 (1872)

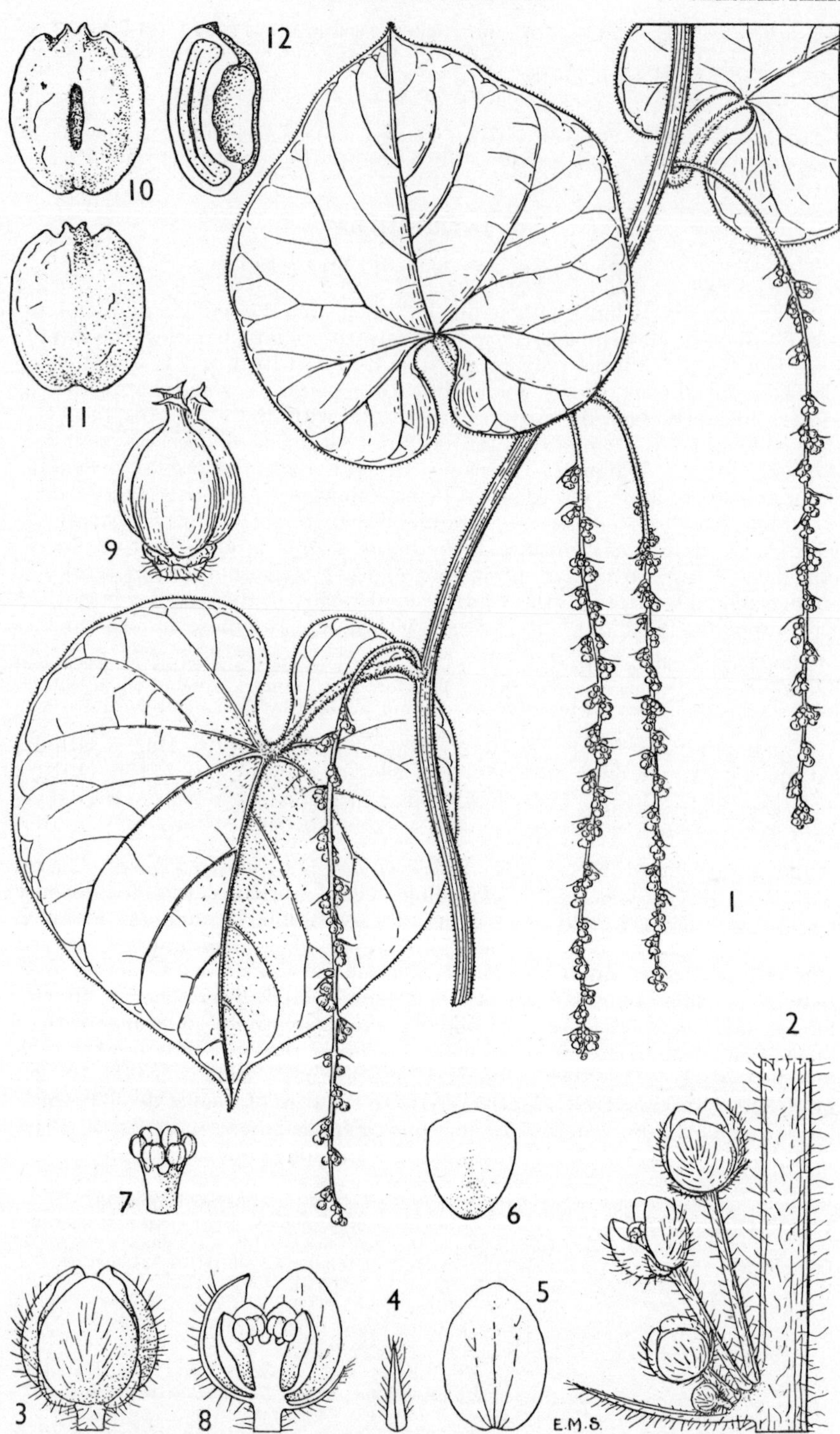

FIG. 5. *CHASMANTHERA DEPENDENS*—**1,** habit, × ⅔ ; **2,** part of ♂ inflorescence, cymule 4-flowered, × 5 ; **3,** ♂ flower, × 10 ; **4,** outer sepal, × 10 ; **5,** inner sepal, × 10 ; **6,** petal, × 10 ; **7,** stamens, × 10 ; **8,** section of ♂ flower, × 10 ; **9,** carpels, × 10 ; **10,** fruit, ventral side, × 2 ; **11,** fruit, dorsal side, × 2 ; **12,** section of seed, × 2. 1–8, from *Keay in F.H.I.* 22871 ; 9, from *Chandler* 2199 ; 10–12, from *Maitland* 691.

7. CHASMANTHERA

Hochst. in Flora 27 : 21 (1844)

Liane with verrucose bark. Leaves simple, longly petiolate, suborbicular-subangular, densely hairy ; nerves palmate. Male inflorescences of false racemes of 3–5-flowered cymules ; bracts filiform, persistent. Male flowers with 6–9 sepals, the 3 outer bract-like and hairy, the 3–6 inner larger, membranous to papery, concave, pubescent outside ; petals 6, fleshy, ribbed inside ; stamens 6, erect, with longly connate filaments ; anther-thecae with longitudinal dehiscence. Female inflorescences of pendulous racemes. Female flowers with sepals and petals similar to those of the ♂, sometimes larger ; staminodes 6, small, elongate ; carpels 3, subovoid, narrowed into a short style ; stigma membranous, recurved, longitudinally cleft. Drupes 3, ellipsoid and unequal-sided, apiculate ; exocarp coriaceous ; endocarp on its dorsal side with a median slightly tuberculate ridge and 3 apical teeth, on its ventral side with 2 narrow marginal wings. Seeds with ruminate endosperm.

Ch. dependens *Hochst.* in Flora 27 : 21 (1844) ; F.T.A. 1 : 41 (1868) ; Diels in E.P. IV. 94 : 152, fig. 51 A–F & J–M (1910) ; Chev., F.V.A.O.F. 1 : 104 (1938) ; F.W.T.A. 1 : 71 (1927) & ed. 2, 1 : 74 (1954) ; U.C.L. : 66 (1935) ; Cufodontis in B.J.B.B. 24, suppl. : 115 (1954). Type : Ethiopia, R. Takkaze, *Schimper* 654 (BR, lecto. !, BM, K, P, S isolecto. !)

Woody liane ; mature branches with flaking bark ; young branchlets densely pubescent. Leaves with petiole 7–14 cm. long ; blade clearly cordate at base, acuminate or subobtuse at apex, 7–20 cm. long and wide, membranous to subpapery, silky-tomentellous when young, later pubescent ; basal nerves 5–7, palmate. Male inflorescences 10–30 cm. long, 1·5–2 cm. wide ; pedicels 3–6 mm. long ; bracts linear-filiform, pubescent or subtomentellous. Male flowers with lanceolate outer sepals 1·5–2 mm. long and 0·5–1 mm. wide ; inner sepals obovate, 2·5–3·5 mm. long, 1·5–2 mm. wide, with a tuft of hairs at apex and sometimes also down the median line outside ; petals subequal, obovate, 2–2·5 mm. long, 1·5–2 mm. wide, glabrous ; stamens 2·5–3 mm. long. Female inflorescences 10–18 cm. long. Female flowers with staminodes about 1 mm. long ; carpels 1·8–2 mm. long, ± united at apex by the stigmas. Drupes 1–1·5 cm. long, 0·8–1·2 cm. wide. Seeds 1–1·8 cm. long. Fig. 5.

UGANDA. Karamoja District : Moroto River, Feb. 1936 (♂ fl.), *Eggeling* 2957 ! ; Entebbe, Nambigiluwa, Feb. 1923 (fr.), *Maitland* 691 !
KENYA. Northern Frontier Province : Dandu, Mar. 1952 (♂ & ♀ fl.), *Gillett* 12622 ! & 12641 ! ; Fort Hall District : Athi River 32 km. SE. of Thika, May 1939 (♂ fl.), *Bally* 9203 !
TANGANYIKA. Ufipa District : Milepa, Feb. 1950 (fr.), *Bullock* 2513 !
DISTR. **U**1–4 ; **K**1, 2, 4 ; **T**4 ; widely spread from Sierra Leone to Somalia ; also in the eastern Belgian Congo
HAB. Lowland rain-forest, riverine forest and, in drier country on termite-hills, near rock-outcrops and in dried-up water-courses, 800–1500 m.

NOTE. *Ch. welwitschii* Troupin of the rain-forests and riverine forests of the Congo Basin is closely related to this species. It differs in the longer inflorescences and the stiffer and fewer hairs. The two species appear to be ecologically distinct.

8. TINOSPORA

Miers in Ann. Mag. Nat. Hist., ser. 2, 7 : 35 (1851), *sensu ampl.*

Desmonema Miers in Ann. Mag. Nat. Hist., ser. 3, 20 : 260 (1867), *non* Rafin. (1833)
Hyalosepalum Troupin in B.J.B.B. 19 : 430 (1949)

Herbaceous or woody lianes, sometimes small, rambling shrubs. Leaves simple, entire ; nerves palmate, rarely pinnate. Male inflorescences of false racemes, which are either simple or compound (i.e. panicles), of 2–4-flowered cymules. Male flowers with 6 sepals, the 3 inner much larger than the outer, hyaline or membranous ; petals 6, rarely 3, fleshy, with inrolled margins, the inner smaller than the outer ; stamens 3–6, quite free, or with their filaments connate either at base, half-way up or for their entire length ; anther-thecae with longitudinal dehiscence. Female inflorescences of false racemes of solitary flowers. Female flowers with sepals and petals similar to those of the ♂ ; carpels 3, obliquely ovoid. Drupes 3 or fewer ; exocarp pulpy ; endocarp ± verrucose outside, with a condyle on its inner side which makes a large subglobular cavity. Seeds with fleshy ruminate endosperm.

Stamens 6, quite free ; inflorescences not exceeding 15 cm. in length ; leaves broadly ovate-triangular, cordate at base . . . . . . 1. *T. bakis*
Stamens 3–6, shortly or longly connate ; inflorescences up to 35 cm. long :
  Leaves subcordate at base, the central part of the base attenuate-cuneate ; stamens 6, with filaments connate at base or to half-way up . 2. *T. mossambicensis*
  Leaves not as above :
    Leaves oblong to oblong-lanceolate, sometimes panduriform, stamens 3 (–6), their filaments connate to apex . . . . . 3. *T. oblongifolia*
    Leaves ovate to suborbicular :
      Stamens 3, their filaments connate to apex ; leaves ovate or ovate-cordate . . 4. *T. caffra*
      Stamens 6, their filaments connate to halfway up ; leaves broadly ovate to suborbicular, normally pale green . . . . . 5. *T. tenera*

1. **T. bakis** (*A. Rich.*) *Miers* in Hook., Niger Fl. 215 (1849) ; Diels in E.P. IV. 94 : 140 (1910). Type : Senegal, *Perrottet* 10 (P, lecto. !)

Liane with branchlets glabrous and verrucose with lenticels. Leaves with sparsely pubescent petioles 0·5–4 cm. long ; blade broadly ovate-triangular, cordate at base, acuminate at apex, 3–5 cm. long and wide, pale green ; basal nerves 5–7, palmate, sometimes puberulous beneath. Male inflorescences 5–12 cm. long ; pedicels 2–3 mm. long. Male flowers with ovate-triangular outer sepals 1·2–1·5 mm. long and 0·4–0·8 mm. wide, inner sepals 2·5–4 mm. long, 1·8–2·8 mm. wide ; petals subequal, 2–3 mm. long, 1·2–1·8 mm. wide ; stamens 6, free, 2·5–3 mm. long. Female inflorescences 7–12 cm. long. Female flowers with carpels 1·2–1·8 mm. long. Drupes 6–9 mm. long, 4–5 mm. wide.

KENYA. Machakos District : Ngomeni, Nov. 1893 (♂ fl.), *Scott Elliot* 6266 !
TANGANYIKA. Mwanza, Feb. 1933, *Rounce* 242 ! & *Wallace* 669 !
DISTR. **K**4 ; **T**1 ; Mauritania, Senegal, French Sudan, Nigeria, Oubangui-Chari, the Sudan and Somalia
HAB. Deciduous bushland and semi-desert scrub, 900–1140 m.

SYN. *Cocculus bakis* A. Rich. in Fl. Seneg. Tent. 1 : 12, t. 4 (1831) ; F.T.A. 1 : 43 (1868) *Chasmanthera bakis* (A. Rich.) Baill., Hist. Pl. 3 : 31 (1872)

2. **T. mossambicensis** *Engl.* in E.J. 26 : 404 (1899). Type : Portuguese East Africa, without locality, *Stuhlmann* 731 (B, holo. !)

Slender liane with glabrous striate branchlets. Leaves with petiole 4–6 cm. long ; blade broadly ovate-triangular, slightly cordate at base, the lateral lobes united in the middle by a projection of the blade towards the petiole giving the central part of the blade a cuneate appearance, glabrous ; nerves 5–7, palmate. Male inflorescences 20–35 cm. long ; pedicels 1·5–2 mm. long. Male flowers with outer sepals 0·8–1·2 mm. long and 1·5–2 mm. wide ; inner sepals 2–2·3 mm. long and 1·5–2 mm. wide ; petals 1·5–2 mm. long, the inner narrower ; stamens 6, their filaments connate at base or to half-way up, 0·8–1·2 mm. long. Female inflorescences 20–35 cm. long ; pedicels 4–5 mm. long. Female flowers with carpels 1·5–2 mm. long. Drupes unknown.

TANGANYIKA. Lindi District : Lake Lutamba, Jan. 1935 (♀ fl.), *Schlieben* 5907 !
DISTR. **T8** ; Portuguese East Africa
HAB. Lowland rain-forest

SYN. *Desmonema mossambicense* (Engl.) Diels in E.P. IV. 94 : 153 (1910) ; Diels in N.B.G.B. 13 : 272 (1936) ; T.T.C.L. : 327 (1949)
Hyalosepalum mossambicense (Engl.) Troupin in B.J.B.B. 19 : 430 (1949)

3. **T. oblongifolia** (*Engl.*) *Troupin* in B.J.B.B. 25 : 137 (1955). Type : Tanganyika, Tanga District, Amboni, *Holst* 2686 (B, holo. !)

Twining liane up to 12 m. high or more ; branchlets glabrous, corky. Leaves with glabrous or very rarely puberulous petioles 1–8 cm. long ; blade oblong or oblanceolate-elliptic, sometimes panduriform, cordate or rounded or rarely slightly sagittate at base, acuminate and mucronulate at apex, 4–10 cm. long, 2–6 cm. wide, glabrous and pale green, thick or somewhat fleshy ; nerves 3–4 pairs. Male inflorescences 10–35 cm. long ; pedicels filiform, 2·5–6 mm. long. Male flowers with triangular outer sepals 0·8–1 mm. long and 1–1·5 mm. wide ; outer petals 1·4–1·8 mm. long and wide, the inner smaller and hardly exceeding 0·5 mm. in width ; stamens 3, rarely 6, their filaments connate to apex. Female inflorescences 20–40 cm. long ; pedicels 4–5 cm. long. Female flowers with carpels 1–1·2 mm. long. Drupes 0·5–0·8 cm. long, 0·4–0·5 cm. wide.

KENYA. Kwale District : Shimoni, Aug. 1953 (♂ fl.), *Drummond & Hemsley* 3905 ! & Shimba Hills, June 1939 (fr.), *van Someren* Sh 115 ! & Buda-Mafisini Forest, near Gazi, Aug. 1953 (♂ fl.), *Drummond & Hemsley* 3798 !
TANGANYIKA. Lushoto District : Makuyuni, June 1935, *Koritschoner* 626 ! ; Tanga District : Kigombe, July 1953 (fr.), *Drummond & Hemsley* 3244 !
ZANZIBAR. Kombeni, Dec. 1930 (♀ fl. & fr.), *Vaughan* 1718 ! ; Haitajwa Hill, Dec. 1930 (♀ fl.), *Greenway* 2656 !
DISTR. **K7**; **T3** ; **Z** ; not known elsewhere
HAB. Lowland rain-forest, coastal evergreen bushland, 0–1000 m.

SYN. *Desmonema oblongifolium* Engl. in E.J. 26 : 408 (1899) ; T.T.C.L. 327 (1949)
Hyalosepalum oblongifolium (Engl.) Troupin in B.J.B.B. 19 : 431 (1949)

4. **T. caffra** (*Miers*) *Troupin* in B.J.B.B. 25 : 137 (1955). Type : South Africa, Natal, *Gerrard* 1976 (K, holo. !, BM, iso. !)

Herbaceous or somewhat woody liane ; stem verrucose with sometimes scaling bark. Leaves with petiole 2·5–10 cm. long ; blade ovate, ovate-cordate or nearly round, cordate or rounded at base, abruptly apiculate and mucronate or subacuminate at apex, 2·5–8 cm. long, 1·5–7 cm. wide, glabrous. Male inflorescences usually of false racemes of cymules, sometimes panicled, 7–30 cm. long ; pedicels 0·3–1 cm. long. Male flowers with triangular to subovate outer sepals 0·8–1·5 mm. long, the inner oblong to oblanceolate, 2–4 mm. long, 1–2 mm. wide ; petals 6, the outer 1·5–3 mm. long, 1–2 mm. wide, the inner narrower ; stamens 3, their filaments connate to apex, 1·5–3 mm. long. Female inflorescences 7–15 cm. long. Petals of ♀ flowers with a small membranous appendage, perhaps a staminode, at base ; carpels 2–3 mm. long. Drupes 0·8–1·2 cm. long, 0·4–0·7 cm. wide.

UGANDA. Ankole District : Bugamba, Apr. 1940 (♂ fl.), *Eggeling* 4224 ! ; Elgon, Busano, Dec. 1936 (♂ fl.), *Snowden* 1034 ! ; Mengo District : Buvuma Island, Mar. 1904, *Bagshawe* 658 !

KENYA. S. Kavirondo District : Kendu-Homa Point, Sept. 1933 (♂ fl.) *Napier* 5272 ! ; Kwale District : Taru, Sept. 1953 (♂ fl.), *Drummond & Hemsley* 4155 !

TANGANYIKA. Ukerewe Island or area Mwanza–Musoma, *Conrads* 5145 ! ; Mpwapwa, Jan. 1935 (♂ fl.), *H. E. Hornby* 616 ! ; Rungwe District : without locality, 1911, *Stolz* 520 !

DISTR. **U**2–4 ; **K**3–5, 7 ; **T**1, 3, 5–7 ; Oubangui-Chari, the Sudan, Belgian Congo, Angola, the Rhodesias, Portuguese East Africa and the Transvaal

HAB. Lowland and upland rain-forest and deciduous bushland, often near rock outcrops, 350–2000 m.

SYN. *Desmonema caffra* Miers in Ann. Mag. Nat. Hist., ser. 3, 20 : 261 (1867) ; Diels in E.P. IV. 94 : 156 (1910)

D. mucronulatum Engl. in E.J. 26 : 409 (1899) ; Uganda Check List : 66 (1935) ; T.T.C.L. : 327 (1949). Type : Tanganyika, Mwanza District, Kagehi [Kayenzi], *Fischer* 70 (B, holo. †) & *Fischer* 69 (B, lecto. !)

D. mucronulatum Engl. var. *schweinfurthii* Engl. in E.J. 26 : 409 (1899) ; Uganda Check List : 66 (1935). Type : Sudan, Equatoria Province, Mt. Baginze [Bangenze], *Schweinfurth* III 70 (B, holo. !)

D. schliebenii Diels in N.B.G.B. 13 : 273 (1936) ; T.T.C.L. : 327 (1949). Type : Tanganyika, Lindi District, Lake Lutamba, *Schlieben* 5838 (B, holo. †, P, lecto. !, G, iso.-lecto. !)

Hyalosepalum caffrum (Miers) Troupin in B.J.B.B. 19 : 431 (1949) ; F.C.B. 2 : 230, tab. 20 (1951)

The species *T. caffra* is here treated in a wide sense—in other words the differences in leaf shape and development of the inflorescence seem not to be of taxonomic importance, but rather due to a somewhat moist and shady habitat. These morphological variations of the plant may well be no more than ecological adaptations of neither specific nor varietal significance.

5. **T. tenera** *Miers* in Contr. Bot. 3 : 37 (1871). Type : Portuguese East Africa, Lower Shire Valley, *Kirk* (K, holo. !)

Liane with yellow-brown glabrous branchlets. Leaves with glabrous petiole 2–3·5 cm. long ; blade broadly ovate to suborbicular, cordate or obtuse at base, acuminate, apiculate and sometimes mucronulate at apex, 4–7 cm. long, 2·5–6 cm. wide, glabrous on both sides, papery, pale green ; nerves 5, palmate. Male inflorescences 10–35 cm. long ; pedicels 2–3 mm. long. Male flowers with oblong-obovate outer sepals 0·5–1 mm. long and 0·4–0·7 mm. wide ; inner sepals obovate-spathulate, 1–1·5 mm. long ; petals keeled, 1–1·5 mm. long ; stamens 6, their filaments connate to halfway up, 1–1·5 mm. long. Female inflorescences, ♀ flowers and fruits unknown.

TANGANYIKA. Morogoro District : Uluguru Mts., Apr. 1933 (♂ fl.), *Schlieben* 3752 ! & Mtibwa Forest Reserve, Dec. 1953 (♂ fl.), *Semsei* 1513 ! & Nov. 1953 (♂ fl.), *Paulo* 178 !

DISTR. **T**6 ; Portuguese East Africa, Northern Rhodesia & the Transvaal

HAB. Lowland rain-forest

SYN. *Tinospora stuhlmannii* Engl. in E.J. 26 : 404 (1899). Type : Portuguese East Africa, Quilimane, *Stuhlmann* I 742 (B, holo. †)

Desmonema tenerum (Miers) Diels in E.P. IV. 94 : 154 (1910) ; T.T.C.L. : 327 (1949)

Hyalosepalum tenerum (Miers) Troupin in B.J.B.B. 19 : 431 (1949)

9. STEPHANIA

Lour., Fl. Cochinch. 2 : 608 (1790)

Ileocarpus Miers in Ann. Mag. Nat. Hist., ser. 2, 7 : 36 (1851), *in clavi*

Homocnemia Miers in Ann. Mag. Nat. Hist., ser. 2, 7 : 36 (1851), *in clavi*

Perichasma Miers in Ann. Mag. Nat. Hist., ser. 3, 18 : 22 (1866) ; Prantl in E. & P. Pf. 3 : 91 (1891)

Herbaceous or woody lianes, sometimes with succulent stems. Leaves simple, peltate ; blade triangular, ovate or suborbicular. Male inflorescences of panicles, false umbels or hemispherical umbel-like cymes. Male flowers with 6–8 free sepals ; petals 3–4, free, rarely absent ; stamens 2–6, arranged in a disc-shaped stalked synandrium, round whose margin the anthers are in a horizontal ring ; thecae with transverse dehiscence. Female inflorescences similar to the ♂. Female flowers with 3–6 free sepals ; petals 2–4, free ; staminodes usually absent ; carpel 1 ; style small ; stigma slightly lobed, or laciniate with divaricate divisions. Drupes with smooth, glabrous or hairy exocarp, endocarp subovate and compressed or reniform, truncate at base, with 2–4 rows of tubercles or ± projecting prickles, or merely with transverse ribs ; the condyle somewhat concave on either side with its septum either perforated or not. Seeds with endosperm.

Stem succulent ; male inflorescences of false umbels of hemispherical cymes arising from leafless stems, very shortly pedunculate ; drupes subreniform ; endocarp transversely ribbed . . . . 1. *S. cyanantha*

Stem woody or herbaceous, not succulent ; inflorescences ± longly pedunculate :

Male inflorescences in false umbels of cymules, axillary, solitary or clustered 2–4 together ; drupes obovate ; endocarp with little prickles or scattered tubercles ; condyle with non-perforated septum . . . 2. *S. abyssinica*

Male inflorescences of elongate panicles arising from leafless stems, rarely axillary ; drupes broadly obovate or suborbicular ; endocarp with 4 prickly ribs, the 2 median ones with prickles enlarged at apex ; condyle with perforated septum . . 3. *S. dinklagei*

1. **S. cyanantha** [*Welw. ex*] *Hiern* in Cat. Afr. Pl. Welw. 1 : 20 (1896) ; Diels in E.P. IV. 94 : 276 (1910) ; Exell & Mendonça in Consp. Fl. Angol. 1 : 42 (1937) ; Troupin in F.C.B. 2 : 244 (1951). Type : Angola, Cuanza Norte, Calinda, *Welwitsch* 2321 (BM, holo. !, K !, LISU, iso.)

Somewhat woody, twining, completely glabrous liane ; stem and branches succulent. Leaves with slender petiole 3–5 cm. long ; blade triangular-orbicular to orbicular, rarely reniform, acuminate to obtuse at apex, entire or with the margins slightly undulate, 4·5–6·5 cm. long and wide, membranous ; basal nerves 10–12, palmate. Male inflorescences of false umbels of subglobular cymes, usually arising from leafless stems, 1–2·5 cm. in diameter ; peduncles and pedicels very short. Male flowers with 6 obovate-incurved, single-nerved sepals 1·5–2 mm. long and 1–1·5 mm. wide ; petals 3–5, broadly subreniform, 0·6–1 mm. long, 1–1·5 mm. wide, fleshy ; synandrium 6–9-locular, 1·2–1·6 mm. in diameter, on a stipe 0·5–1 mm. long. Female inflorescences similar to the ♂. Female flowers unknown. Drupes subreniform, 4–6 mm. long ; exocarp like a thin skin, red when alive ; endocarp transversely ribbed and furrowed. Seeds 5–7 mm. long.

Kenya. Elgon, Jan. 1931 (♂ fl.), *Lugard* 502 ! & Apr. 1931 (fr.), *Lugard* 502a !
Tanganyika. Rungwe District : without locality, 1913, *Stolz* 2241 !
Distr. **K5** ; **T7** ; Fernando Po, French Cameroons, Belgian Congo, Angola and Northern Rhodesia
Hab. This rather rare species shows a preference for the higher altitudes up to 2200 m., especially in swamps or on volcanic lava. Its precise ecological requirements are still unknown. According to Welwitsch this species is an epiphyte !

According to Welwitsch also, this species is monoecious. This character—a rare one in the family—requires to be verified. The androecium is quite typical and characteristic of *Stephania*, but the inflorescences and the fruit are very different from those of the other species at present placed in this genus.

2. **S. abyssinica** (*Dillon & A. Rich.*) *Walp.*, Rep. 1 : 96 (1842). Type : Ethiopia, Adowa, *Quartin-Dillon & Petit* (P, holo., K, iso. !)

Twining liane, woody at base ; stem covered with a thin bark ; branchlets glabrous or ± densely pubescent to tomentose when young. Leaves with petiole 4–12 cm. long ; blade ovate to broadly ovate, rarely suborbicular, rounded at base, obtuse or subacute at apex, 5–20 cm. long, 4–13 cm. wide, membranous or papery, slightly discolorous, glabrous or tomentellous ; basal nerves 8–10, palmate. Male inflorescences of false compound umbels, axillary, solitary or clustered 2–4 together ; axes glabrous or tomentellous ; peduncle 4–10 cm. long, with 3–6 rays ending in umbel-like cymes ; involucre of 3–5 caducous bracts. Male flowers with 6–8 obovate or subobovate sepals 1·2–2·5 mm. long and 0·6–1·2 mm. wide, purplish, their base often violet ; petals 3–4, broadly ovate or suborbicular, 0·8–1·2 mm. long ; synandrium 6–8-locular. Female inflorescences similar to the ♂. Female flowers with 3–4 sepals ; carpel glabrous. Drupes subspherical-flattened, 0·5–0·8 cm. in diameter, glabrous ; endocarp with small prickles or thick tubercles arranged in three lines ; condyle not perforated. Seeds 1·5–2 cm. long.

HAB. A species with varied ecological requirements, not however penetrating into the rain-forest ; in grassland or wooded grassland up to 3500 m., preferably in moist shady places, especially on the edges of rivers and swamps.

var. **abyssinica**

Young branchlets, petioles, lower sides of leaves, inflorescences and (partly) the sepals glabrous.

UGANDA. West Nile District : Zeio, Mar. 1935 (♂ fl.), *Eggeling* 1894 ; Ruwenzori, Nanwamba Valley, Jan. 1935, *G. Taylor* 3063a ! ; Elgon, above Butandiga, Jan. 1918 (♂ fl.) *Dummer* 3641 !

KENYA. Aberdare Mts., Jan. 1922 (♀ fl.) *Fries* 713 ! ; Machakos District : Chyulu Hills, Apr. 1938, *Bally* 8006 ! ; Kericho District : Sotik, Kibajet Estate, Nov. 1949 (♂ fl.), *Bally* 7445a !

TANGANYIKA. Morogoro, Uluguru Mts., Feb. 1933 (♂ fl.), *Schlieben* 3443 ! ; W. Usambara Mts., near Makuyuni, June 1935 (♀ fl. & fr.), *Koritschoner* 826 ! ; Rungwe District : Mt. Rungwe, Sep. 1912 (♀ fl. & fr.), *Stolz* 1558 !

DISTR. **U**1–4 ; **K**3–5 ; **T**2, 3, 6, 7 ; widely spread from French Guinea east to Ethiopia and south through Belgian Congo to Angola, Basutoland & Natal

SYN. *Stephania abyssinica* (Dillon & A. Rich.) Walp., Rep. 1 : 92 (1842) ; F.T.A. 1 : 47 (1868) ; Diels in E.P. IV. 94 : 268, fig. 89 (1910) ; F.W.T.A. 1 : 74 (1927) ; Uganda Check List : 66 (1935), *pro parte* ; T.T.C.L. : 328 (1949) ; Troupin in F.C.B. 2 : 245 (1951) ; Cufodontis in B.J.B.B. 24, suppl. : 114 (1954)
Clypea abyssinica Dillon & A. Rich. in Ann. Sc. Nat., sér. 2, 14 : 38 (1840)
Ileocarpus schimperi Miers in Ann. Mag. Hist., ser. 3, 14 : 373 (1864). Type : Ethiopia, Tigré, *Schimper* 178 (BM, holo. !, K, iso. !)

var. **tomentella** (*Oliv.*) Diels in E.P. IV. 94 : 270 (1910) ; Uganda Check List. : 66 (1935) ; T.T.C.L. : 328 (1949) ; Troupin in F.C.B. 2 : 246 (1951) ; Cufodontis in B.J.B.B. 24, suppl. : 114 (1954) ; F.W.T.A., ed. 2, 1 : 75 (1954). Type : Tanganyika, Kilimanjaro, *Johnston* (K, holo. !, BM, iso. !)

Young branchlets, petioles, lower sides of leaves, inflorescences and (partly) the sepals ± densely pubescent to tomentose.

UGANDA. Kigezi District : Kachwekano, Dec. 1949 (fl. buds), *Purseglove* 3135 ! ; Mengo District : Kawanda, *Hazel* 392 !

KENYA. Elgon, Dec. 1930 (♂ fl.), *Lugard* 447 ! ; Kiambu District : Limuru, Feb. 1915 (♀ fl.), *Dummer* 1599 !

TANGANYIKA. Kilimanjaro, Marangu, June 1927 (♀ fl.), *Haarer* 515 ! & Lyamungu, Aug. 1932 (♂ fl.), *Greenway* 3102 ! ; Morogoro District : Uluguru Mts., above Bunduki, Mar. 1953 (♂ fl.), *Drummond & Hemsley* 1493 !

DISTR. **U**2–4 ; **K**2–5 ; **T**2, 3, 6 ; widely spread from the British Cameroons east to Ethiopia and south to the Cape Province

SYN. *Stephania praelata* Miers in Contr. Bot., 3 : 230 (1871) ; Uganda Check List : 66 (1935). Type : South Africa, Orange Free State, *Cooper* 904 (BM, holo. !, K, iso. !)

S. hernandifolia (Willd.) Walp. var. *tomentella* Oliv. in Trans. Linn. Soc., ser. 2, 2 : 328 (1887)
S. hernandifolia (Willd.) Walp. var. *pubescens* Szyszyl., Polyp. Thal. Rehm.: 102 (1887). Type : South Africa, Natal, Drakensberg, Coldstream, *Rehmann* 6895 (K, iso.-syn. !)
S. abyssinica (Dillon & A. Rich.) Walp. var. *pubescens* Engl. in P.O.A. C : 181 (1895). Type : Tanganyika, Moshi District, Kibosho, *Volkens* 1601 (B, holo. †)
[*S. hernandifolia* sensu Uganda Check List : 66 (1935), *non* (Willd.) Walp.]
[*S. abyssinica* sensu Uganda Check List : 66 (1935), *pro parte, quoad spec. Maitland* 1221 & *Snowden* 892, *non* (Dillon & A. Rich.) Walp.]

3. **S. dinklagei** (*Engl.*) Diels in E.P. IV, 94 : 265 (1910) ; F.W.T.A. 1 : 75 (1927) & ed. 2, 1 : 75 (1954) ; Chev., F.V.A.O.F. 1 : 126 (1938) ; Troupin in F.C.B. 2 : 245 (1951). Type : French Cameroons, Grand Batanga, *Dinklage* 943 (B, holo. !)

Liane with woody stem, sometimes a rambling shrub ; branchlets covered with greyish-brown glabrous bark. Leaves with glabrous petiole 6–12 cm. long ; blade broadly ovate to suborbicular, rounded at base, acuminate at apex, 7–15 cm. long and wide, membranous to papery, discolorous ; basal nerves 8–10, palmate. Male inflorescences of long panicles of cymules up to 50 cm. long, generally arising from the leafless stems, rarely axillary ; cymules ± umbel-like, 2–5 cm. long ; bracts asymmetric, up to 2·5 cm. long and 1·5 cm. wide ; pedicels 1 mm. long. Male flowers with narrowly obovate and somewhat incurved pubescent sepals 0·9–1·8 mm. long and 0·6–0·8 mm. wide ; petals 3, broadly ovate or obtriangular, 0·4–0·6 mm. long and wide ; synandrium 6-locular, 0·7–0·9 mm. long. Female inflorescences probably similar to the ♂, but shorter. Female flowers unknown. Drupes obovate, subtruncate at base, 0·6–1·2 cm. long ; endocarp with 4 prickly ribs, the two median ribs broadened at apex ; condyle with perforated septum. Seeds 0·8–1 cm. long.

UGANDA. Kigezi District : Ishasha Gorge, May 1950 (♂ fl.), *Purseglove* 3432 !
TANGANYIKA. Ukerewe Is. or area Mwanza–Musoma, *Conrads* 5816 ! & *Conrads in E.A.H.* 10428 !
DISTR. **U2** ; **T1** ; in western Africa from French Guinea to Cabinda & Belgian Congo
HAB. Lowland rain-forest and riverine forest, 1100–1500 m.

SYN. *Cissampelos dinklagei* Engl. in E.J. 26 : (1899)
Stephania dinklagei (Engl.) Diels var. *axillaris* Troupin in B.J.B.B. 19 : 434 (1949) ; F.C.B. 2 : 245 (1951). Type : Belgian Congo, Yangambi, *Louis* 1753 (BR, holo. !)
Kolobopetalum tisserantii Chev., F.V.A.O.F. 1 : 117 (1938) *gallice, pro parte, quoad inflorescentia.* Type : Oubangui-Chari, Bambari, *Tisserant* 1525 (P, holo. !)
[*Stephania abyssinica* sensu Diels in E.P. IV. 94 : 268 (1910), *pro parte, quoad spec. Jolly* 155 (Gabon), *non* (Engl.) Diels.]

10. CISSAMPELOS

L., Sp. Pl. 1031 (1753) ; Gen. Pl., ed. 5, 1138 (1754), *pro parte*

Twining lianes, sometimes rambling shrubs. Leaves simple, peltate or subpeltate, entire or angular. Female inflorescences of corymbose cymules which are either solitary or clustered or arranged in ± developed false racemes. Male flowers with 4 (–5) obovate often spreading sepals ; petals usually connate into a patelliform or cup-shaped corolla, sometimes incompletely connate ; stamens connate into a 4–10-locular synandrium ; anther-thecae with longitudinal dehiscence. Female inflorescences less developed than the ♂ ; cymules 3–9-flowered, axillary or in false racemes, arising from the axils of leaves or of accrescent bracts. Female flowers with 1 sepal, rarely more ; petal 1, rarely 2 or 3, smaller than the sepal, sometimes broader than long ; carpel 1. Drupes with hairy or glabrous exocarp ;

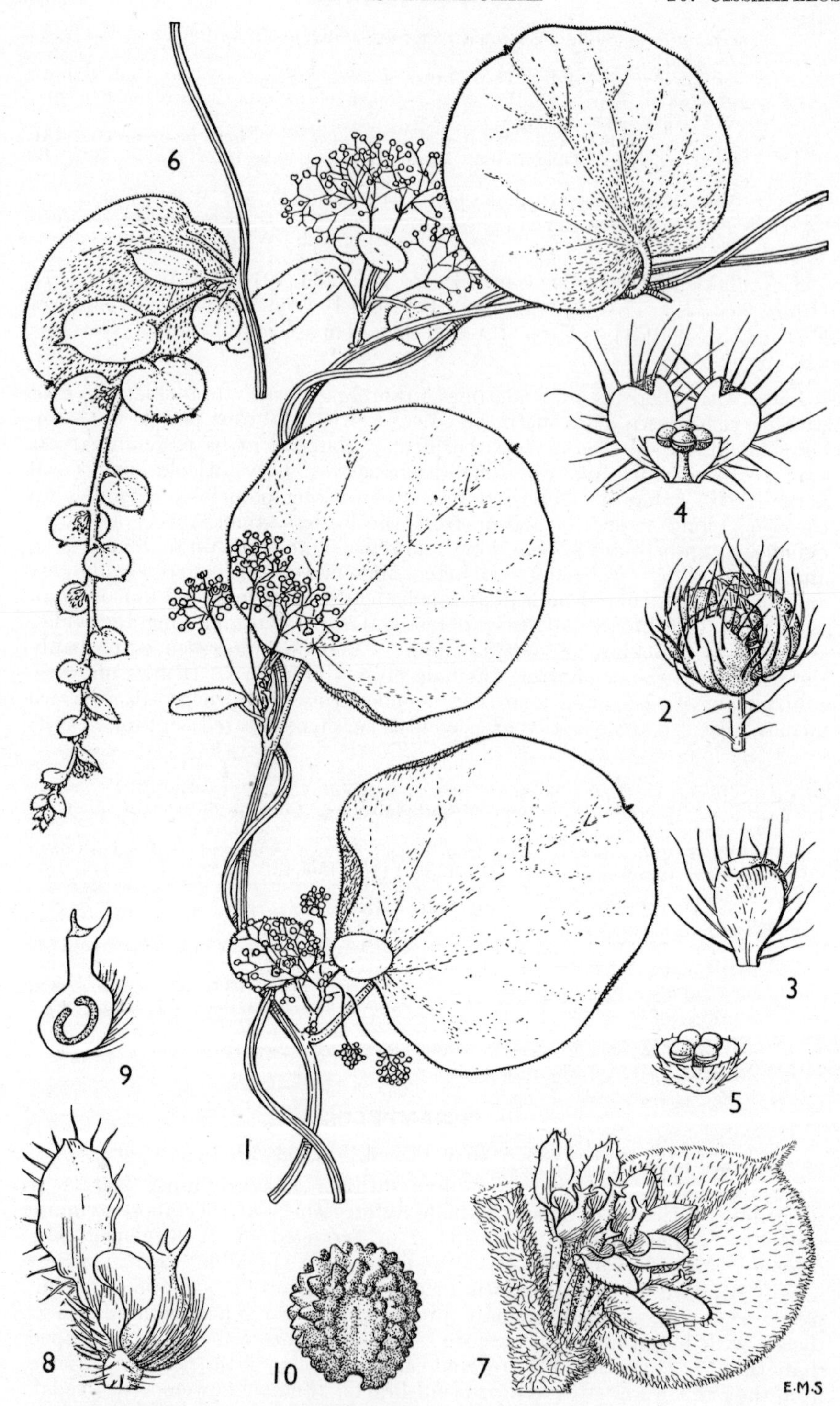

FIG. 6. *CISSAMPELOS PAREIRA* var. *ORBICULATA*—**1,** habit of ♂ plant, × **1** ; **2,** ♂ flower, × 20 ; **3,** sepal, × 10 ; **4,** section of flower, × 20 ; **5,** corolla and stamens, × 20 ; **6,** habit of ♀ plant, × 1 ; **7,** part of ♀ inflorescence, cymule and bracts, × 10 ; **8,** ♀ flower, × 20 ; **9,** ovary, × 20 ; **10,** seed, × 5. 1, 6–10, from *Graham* 1535, 2–5, from *Bax* 2.

mesocarp fleshy and thin ; endocarp woody, with 1 dorsal ridge ; the sides with small transverse often verrucose ribs. Seeds with a sparse endosperm.

Leaves subpeltate :
- Leaves rounded to emarginate or sometimes obtuse at apex :
  - Branchlets sparsely hairy (hairs ± spreading to glabrescent) ; leaves broadly ovate-triangular, sometimes subreniform, usually truncate or somewhat rounded at base, obtuse or somewhat rounded at apex . . . . 1. *C. truncata*
  - Branchlets pubescent to puberulous (hairs crisped and subappressed) :
    - Leaves normally suborbicular, rounded or sometimes subcordate at base, rounded at apex ; ♂ inflorescences of fascicled and axillary corymbose cymules, rarely of cymules arranged along an axis not exceeding 10 cm. in length ; synandrium 4-locular . . 2. *C. pareira* var. *orbiculata*
    - Leaves heart-shaped to subreniform, obtuse to somewhat rounded at apex, ± deeply cordate at base ; ♂ inflorescences usually of cymules arranged along an axis up to 15 cm. long ; synandrium 6–8-locular . . . . . 3. *C. mucronata*
- Leaves clearly acuminate at apex, ovate-elliptic ; branchlets, petioles and nerves on the lower side of the leaves densely pubescent ; inflorescences not exceeding 1·5 cm. long . . . . . 4. *C. friesiorum*

Leaves clearly peltate :
- Bracts of the ♀ inflorescences and of the infructescences ± longly acuminate ; leaves truncate to subcordate at base, elongate-acute at apex, petiole densely tawny-pubescent . . . . 5. *C. insignis*
- Bracts of the ♀ inflorescences and of the infructescences suborbicular to subreniform :
  - Petiole of adult leaves 4–16 cm. long ; blade generally membranous to somewhat papery :
    - Plants usually hairy-pubescent ; leaves generally pentagonal or angular, sometimes triangular to broadly ovate, obtuse or acute or emarginate and mucronulate at apex ; ♂ inflorescences up to 40 cm. long . . . . . 6. *C. owariensis*
    - Plants normally sparingly puberulous to glabrescent ; leaves suborbicular, rounded and emarginate at apex ; ♂ inflorescences not exceeding 5 cm. in length . . . . . 7. *C. nigrescens*
  - Petiole of the adult leaf 1–3·5 (–5) cm. long ; blade papery or subcoriaceous :
    - Leaves not orbicular, heart-shaped to pentagonal, densely puberulous or tomentellous beneath ; ♂ inflorescences up to 30 cm. long . . . 8. *C. rigidifolia*
    - Leaves orbicular, puberulous beneath ; ♂ inflorescences not exceeding 5 cm. in length 2. *C. pareira* var. *orbiculata*

1. **C. truncata** *Engl.* in E.J. 26 : 398 (1899) ; Diels in E.P. IV. 94 : 297 (1910) ; T.T.C.L. : 326 (1949). Type : Tanganyika, Uluguru Mts., *Stuhlmann* 8831 (B, holo. !)

Slender, somewhat woody liane ; branchlets sparsely clothed with caducous yellowish-white hairs. Leaves with slender petioles 2·5–8 cm. long, sometimes ± densely whitish-pubescent towards apex, inserted 2 mm. at most from the base of the blade ; blade ovate-triangular or sometimes reniform, truncate or somewhat rounded at base, obtuse to somewhat rounded and mucronate at apex, 2·5–6 cm. long, 3–8 cm. wide, membranous, very sparingly pubescent to glabrescent above, slightly pubescent at the base of the nerves beneath ; basal nerves 3–5, palmate. Male inflorescences of axillary, generally solitary or paired, corymbose cymules not exceeding 5 cm. in length, or of cymules arranged in the axils of membranous bracts along an axis up to 10 cm. in length ; peduncles 1–4 cm. long, filiform. Male flowers with glabrous sepals 1–1·5 mm. long ; corolla cup-shaped, glabrous, sometimes incompletely connate, 0·7–1 mm. long ; synandrium (5–) 6–8-locular, 0·8–1 mm. long. Female inflorescences, ♀ flowers and fruits unknown.

TANGANYIKA. E. Usambara Mts., Bomole, July 1917 (♂ fl.), *Zimmermann in A.H.* 6405 ! ; Ufipa District : Mbisi, Apr. 1950 (♂ fl.), *Bullock* 2816 ! ; Morogoro District : Uluguru Mts., Oct. 1934 (♂ fl.), *E. M. Bruce* 74 !
DISTR. **T3**, 4, 6 ; known only from Tanganyika
HAB. Lowland and upland rain-forest and mist-forest, 900–2300 m.

NOTE. The specimen, *Grote* 3842 (Tanganyika, Amani), is placed under this species with doubt, on account of its small leaves, acuminate at apex and not exceeding 1·5 cm. in length.

2. **C. pareira** *L.* var. **orbiculata** (*DC.*) *Miquel* in Ann. Mus. Lugd. Bat. 4 : 85 (1868) ; Troupin in B.J.B.B. 25 : 140 (1955). Type : East Indies, locality uncertain, *Roxburgh* (G–DC, holo. !)

Liane with stem somewhat woody at base. Leaves with puberulous to subtomentose petioles 1–7 cm. long and inserted 1–4 mm. from the base of the blade ; blade suborbicular or broadly ovate, rounded or subcordate or subtruncate at base, rounded or emarginate and mucronulate at apex, 2–12 cm. long and wide, membranous to papery, densely puberulous to tomentose beneath ; basal nerves 5–7, palmate. Male inflorescences of axillary, generally solitary or paired, corymbose cymules not exceeding 5 cm. in length, sometimes arranged in the axils of bracts along an axis up to 10 cm. long ; axes, peduncles and pedicels whitish-pubescent. Male flowers with 4–5 ovate or obovate keeled sepals 1·2–1·5 mm. long and 0·7 mm. wide, tubercled and hairy-pubescent outside ; corolla cup-shaped, 0·7–1 mm. long, sparsely pubescent ; synandrium 4-locular. Female inflorescences of 5–9-flowered cymules arranged in axillary false racemes 5–10 cm. long, solitary or clustered 2–3 together ; bracts suborbicular-reniform, up to 1·5 cm. in diameter, pubescent-tomentose. Female flowers with sepals similar to those of the ♂ ; petal obtriangular to subreniform, 1·5–1·7 mm. long, 2 mm. wide, very sparsely pubescent. Drupes 4–6 mm. long, 3–4 mm. wide, hairy-pubescent. Fig. 6, p. 24.

UGANDA. Kigezi District : Kachwekano, May 1951 (♀ fl.), *Purseglove* 3616 ! ; Mbale District : Bugishu, Babungi, July 1926 (♀ fl. & fr.), *Maitland* 1234 !
KENYA. Machakos District : Kibwezi, Mar. 1906 (♀ fl.), *Scheffler* 147 ! ; Nairobi, Dec. 1937 (♂ fl.), *van Someren* 1521 ! ; Kilifi District : Kibarani, Oct. 1945 (♀ fl. & fr.), *Jeffery* K360 !
TANGANYIKA. Shinyanga, *Koritschoner* 1810 ! ; Lushoto District : Mkuzi, Apr. 1953 (♀ fl. & fr.), *Drummond & Hemsley* 2115 ! ; Morogoro, Kiroka, Oct. 1932 (♀ fl.), *Schlieben* 2888 !
ZANZIBAR. Zanzibar Is., Mangapwani, Dec. 1930 (♂ fl.), *Greenway* 2625 ! & Haitajwa, Sep. 1930 (♂ fl.), *Vaughan* 1567 !

DISTR. **U**2, 3 ; **K**1–4, 7 ; **T**1–3, 5, 6 ; **Z** ; **P** (*fide* U.O.P.Z. : 192 (1949)—no specimen seen) ; Ethiopia, Southern Rhodesia ; also through tropical Asia to the East Indies

HAB. Upland and lowland rain-forest, coastal evergreen bushland and deciduous bushland, often persisting on cleared ground and in plantations ; also in secondary vegetation and near rock outcrops, 0–2300 m.

SYN. *C. orbiculata* DC., Syst. 1 : 537 (1817)
C. pareira L. var. *mucronata* (A. Rich.) Engl. subvar. *crassifolia* Engl. in E.J. 26 : 394 (1899), *pro parte*
C. pareira L. var. *mucronata* (A. Rich.) Engl. subvar. *usambarensis* Engl. in E.J. 26 : 395 (1899). Type : Tanganyika, Usambara Mts., *Buchwald* 637 (B, holo. †, K, iso. !)
C. pareira L. var. *transitoria* Engl. subvar. *wakefieldii* Engl. in E.J. 26 : 396 (1899). Type : Kenya, Mombasa, *Wakefield* (B, holo.†)
C. pareira L. var. *typica* Diels in E.P. IV. 94 : 288 (1910), *non* var. *pareira* sensu L.
[*C. pareira* sensu *auct. afric.*, *non* L.]

VARIATION. The characters of the var. *orbiculata*, even in the restricted sense adopted above, which contrasts with the treatment by Diels, are not entirely homogeneous. Several variations may be found : the progressive disappearance of the indumentum in specimens coming from a shaded habitat (e.g. *Greenway* 2625), or an increase in the density of the indumentum in specimens collected at high altitudes (*Andersen*). The subreniform shape of the leaves seen in certain specimens (e.g. *Bally* 1888) recalls that of *C. truncata* E. Mey. This last species may be separated vegetatively from *C. pareira* var. *orbiculata* by the less robust habit of the plant and also by the less compact ♂ inflorescences ; moreover, the indumentum of *C. truncata*, when it is present, is composed of yellowish and much larger hairs.

3. **C. mucronata** *A. Rich.* in Fl. Seneg. Tent. 1 : 11 (1831) ; Diels in E.P. IV. 94 : 300 (1910) ; F.W.T.A. 1 : 75 (1927) & ed. 2, 1 : 75 (1954) ; T.T.C.L. 326 (1949) ; Troupin in F.C.B. 2 : 250, tab. 22 (1951) ; Cufodontis in B.J.B.B. 24, suppl. : 114 (1954). Type : Senegal, Walo, 1830, *Perrottet* (P, lecto. !, BM, iso. !)

Liane with woody rootstock ; branchlets hairy-pubescent to subtomentose. Leaves subpeltate, with petioles 2–4·5 cm. long and inserted 0·5–3 mm. above the base of the blade ; blade ovate-heartshaped, broadly or narrowly cordate at base, normally obtuse or somewhat rounded and mucronulate at apex, 4–12 cm. long, 4–13 cm. wide, tomentellous or puberulous or glabrescent on both sides with a bright grey or yellowish indumentum ; basal nerves 5–7, palmate. Male inflorescences of corymbose cymules clustered 3–6 together and either axillary or arranged in false racemes 5–15 cm. long ; pedicels 1–2 mm. long. Male flowers with 4–5 sepals 1–1·5 mm. long and 0·7–1 mm. wide, hairy-pubescent outside ; corolla cup-shaped, 1–1·5 mm. long and in diameter, spreading after flowering ; synandrium 6–10-locular, 1–1·5 mm. long. Female inflorescences 5–16 cm. long ; bracts accrescent, often longly mucronulate. Female flowers with sepal 1·5 mm. long and 1 mm. wide, pubescent ; petal 0·7 mm. long, 2 mm. wide, glabrous ; carpel 1–1·2 mm. long, glabrescent. Drupes 0·4–0·7 mm. long, 0·3–0·5 mm. wide, pubescent.

UGANDA. Bunyoro District : Mutunda, Mar. 1943 (♀ fl. & fr.), *Purseglove* 1323 ! ; Kutwi between Entebbe and Butiaba, Dec. 1909, *Mearns* ! ; Mengo District : Kyagwe, Namanve Swamp, Mar. 1952 (♀ fl. & fr.), *Eggeling* 210 *in F.D.* 465 !

KENYA. Uasin Gishu District : Kipkarren, Dec. 1931 (fl. buds), *Brodhurst-Hill* 662 ! ; Kisumu, Feb. 1915 (♀ fl. & fr.), *Dummer* 1789 ! & 1854 ! ; Masai District : Mara Masai Reserve, Telek river, Sep. 1947 (♂ fl.), *Bally* 5318 !

TANGANYIKA. Singida District : Iwumbu River, Aug. 1937 (♂ fl.), *B. D. Burtt* 767 ! ; Morogoro District : Uluguru Mts. Bahati, Apr. 1935 (♂ fl.), *E. M. Bruce* 1053 ! ; Iringa, June 1936 (♂ fl.), *Emson* 535 !

ZANZIBAR. Mwero Swamp, Oct. 1930 (♂ fl.), *Vaughan* 1604 !

DISTR. **U**1–4 ; **K**3, 5, 6 ; **T**1–3, 5–7 ; **Z** ; from Senegal eastwards to Ethiopia and southwards to South West Africa, the Transvaal and Natal

HAB. Deciduous bushland, often on termite-hills and near rock-outcrops, riverine forest and in swamps ; often persists on cultivated land, 0–1800 m.

SYN. *C. pareira* L. var. *mucronata* (A. Rich.) Engl. in P.O.A. C : 181 (1895) ; Dur. & Schinz, Consp. Fl. Afr. 1(2) : 51 (1898)
C. pareira L. var. *mucronata* (A. Rich.) Engl. subvar. *crassifolia* Engl. in E.J. 26 : 394 (1899), *pro parte*
C. pareira L. var. *pachyphylla* Diels in E.P. IV : 301 (1910) ; T.T.C.L. : 326 (1949)
C. pareira L. var. *crassifolia* (Engl.) Engl. in Z.A.E. : 211 (1911)
[*C. pareira* sensu Oliv., F.T.A. 1 : 46 (1868), *pro parte, non* L.]

4. **C. friesiorum** *Diels* in N.B.G.B. 8 : 477 (1923). Type : Kenya, Meru, *Fries* 1625 (S, holo. !)

Slender liane ; branchlets clothed with pale yellow hairs. Leaves with petioles 2–3 mm. long, clothed with pale yellow pubescence and inserted 1–3 mm. from the base of the blade ; blade ovate, sometimes asymmetric, somewhat rounded at base, longly acute and mucronulate at apex, slightly sinuate on the margins, 3·5–7 cm. long, 2·5–5 cm. wide, membranous, pubescent above, subtomentose beneath ; basal nerves 5–7, palmate. Male inflorescences of axillary cymules clustered 2–3 together ; peduncle 1–1·2 cm. long, pubescent ; pedicels 3–4 mm. long. Male flowers with sepals 1–1·2 mm. long and 0·8–1 mm. wide, hairy-pubescent outside ; corolla cup-shaped, about 1 mm. long, glabrous ; synandrium 4-locular, 0·8–1 mm. long. Female inflorescences, ♀ flowers and fruits unknown.

KENYA. Meru, Feb. 1922 (♂ fl.), *Fries* 1625 !
DISTR. **K4** ; not known elsewhere
HAB. Upland rain-forest, about 2000 m.

NOTE. The specimen *Conrads in E.A.H.* 10434 (Tanganyika, Ukerewe Is.) is near this species in the shape of its leaves ; it differs however in the usually less dense and shorter pubescence, in the more shortly acute leaves, and in the more developed inflorescences. The specimen may perhaps fall within the range of variation of *C. pareira* L. var. *orbiculata* (DC.) Miquel.

5. **C. insignis** *Alston* in K.B. 1925 : 362 (1925) ; T.T.C.L. : 325 (1949). Type : Tanganyika, Rungwe District, Kyimbila gorge, *Stolz* 1600 (K, holo. !, BM, BR, IFI, PRE, UPS, iso. !)

Twining liane ; branchlets with long yellow pubescence. Leaves with petioles 2·5–3·5 cm. long densely pubescent with yellow-tawny hairs and inserted 1–1·5 cm. from the base of the blade ; blade broadly ovate, truncate to subcordate at base, longly acute to mucronate at apex with margins entire or slightly sinuate, 10–14 cm. long, 7·5–11 cm. wide, sparsely pubescent to glabrescent above, pubescent beneath especially on the nerves, densely ciliolate ; basal nerves 7–11, palmate. Male inflorescences and flowers unknown. Female inflorescences 10–13 cm. long ; bracts subsessile, broadly ovate, sometimes subpeltate, acuminate and acute at apex, 1·5–2 cm. long, 1–1·8 cm. wide, pubescent and densely ciliolate ; pedicels 1–1·5 cm. long. Female flowers with 1 or rarely 2 sepals 0·8–1 mm. long ; petal transversely elliptic, slightly pubescent ; carpel about 1 mm. long ; ovary pubescent. Drupes 4·5–7 mm. long, 4–5 mm. wide, pubescent. Seeds 6–8 mm. long.

TANGANYIKA. Rungwe District : Kyimbila gorge, *Stolz* 1600 !
DISTR. **T7** ; not known elsewhere
HAB. Unknown

6. **C. owariensis** [*Beauv. ex*] *DC.*, Prodr. 1 : 100 (1824) ; Diels in E.P. IV. 94 : 302 (1910) ; F.W.T.A. 1 : 75 (1927) & ed. 2, 1 : 75 (1954) ; T.T.C.L. : 326 (1949) ; Troupin in F.C.B. 2 : 249 (1951), *pro parte.* Type : Nigeria, Owari [Warri] *Beauvois* (G–DC, holo. !, BM, iso. !)

Liane with stem and branchlets ± densely hairy-pubescent, rarely hairy and puberulous, sometimes ultimately glabrescent ; indumentum normally composed of spreading hairs. Leaves with petiole 4–16 cm. long inserted 0·8–2 cm. from the base of the blade ; blade broadly ovate, generally suborbicular and angular (with the angles more or less prominent) or broadly triangular, truncate or rounded or subcordate at base, obtuse and mucronulate at apex, with variable indumentum, very rarely glabrous beneath ; basal nerves 5–7, palmate. Male inflorescences up to 40 cm. long ; cymules with pubescent peduncles 0·5–3 cm. long. Male flowers with sepals 1–1·5 mm. long and 0·7–1·2 mm. wide, hairy outside ; corolla cup-shaped, 1–1·2 mm. long. Female inflorescences up to 35 cm. long ; bracts suborbicular or reniform, up to 4 cm. in diameter, mucronulate, hairy, longly ciliolate ; pedicels 1–1·2 mm. long. Female flowers with sepal 1·2–2 mm. long and 0·7–1 mm. wide ; petal truncate or subreniform, 1–1·2 mm. long and wide ; carpel 1–1·3 mm. long, hairy. Drupes 4–6 mm. long, 4–5 mm. wide, hairy.

TANGANYIKA. Ulanga District : Masagati, June 1931 (fr.), *Schlieben* 1096 ! ; Rungwe District : Lupata, *Davies* 252 ! ; Lindi District : Rondo Plateau, Mchinjiri, Mar. 1952 (fr.), *Semsei* 686 !

DISTR. **T**6–8 ; from Sierra Leone to the Belgian Congo, Angola and Northern Rhodesia

HAB. Lowland rain-forest and riverine forest, up to 900 m.

SYN. *C. insolita* [Miers ex] Oliv., F.T.A. 1 : 46 (1868). Type : Gabon, Corisco Bay, *Mann* 1870 (K, holo. !)
C. pareira L. var. *owariensis* (Beauv. ex DC.) Oliv., F.T.A. 1 : 46 (1868)
C. pareira L. subsp. *owariensis* (Beauv. ex DC.) Engl. in E.J. 26 : 396 (1899)
C. robertsonii Exell in J.B. 64 : 192 (1936). Type : Togo, Kpandu, *Robertson* 27 (BM, holo. !)

NOTE. The Tanganyika specimens are untypical, which, however is not surprising since they come from near the limit of the geographical range of the species. *Schlieben* 1096 has been annotated in the herbarium as the type of *C. ciliosa* Peter, a species not however described ; the leaves are unusually big, especially in the specimen at Stockholm, and without angles.

7. **C. nigrescens** *Diels* in E.P. IV. 94 : 296 (1910). Type : Tanganyika, E. Usambara Mts., Amani, *Warnecke* 446 (B, holo. !, BM, EA, K, P, iso. !)

Liane with sparingly pubescent to glabrescent branchlets. Leaves with sparingly pubescent to glabrescent petioles 2·5–16 cm. long and inserted 0·4–1 cm. from the base of the blade ; blade suborbicular, sometimes subreniform, truncate or subcordate to deeply cordate at base, rounded to subemarginate or sometimes obtuse or deeply emarginate at apex, 3·5–8 cm. long, 4–11 cm. wide, membranous to papery, blackish, glabrescent and somewhat glossy above. Male inflorescences 3·5–5 cm. long ; axes sparingly pubescent ; pedicels 0·5–1 mm. long. Male flowers with oblong-obovate sparingly puberulous sepals 2 mm. long and 1 mm. wide ; corolla cup-shaped, 0·8–1 mm. long ; synandrium 4-locular, 0·8–1 mm. long. Female inflorescences 5–20 cm. long ; bracts subreniform, densely rusty-ciliolate, up to 3·5 cm. long. Female flowers with sepal 2–2·5 mm. long and 0·8–1 mm. wide, with some brown hairs ; petal suborbicular to truncate at apex, sometimes subreniform, 1 mm. long and wide, sparingly hairy-pubescent ; carpel 1–1·2 mm. long, longly hairy. Drupes 3–5 mm. long, 2–4 mm. wide, hairy-pubescent.

var. **nigrescens**

Leaves suborbicular, sometimes subcordate at base, rounded and mucronulate at apex, not exceeding 9 cm. in width. Male inflorescences and flowers unknown.

TANGANYIKA. E. Usambara Mts., Nguelo [Ngwelo], Oct. 1916 (fr.), *Zimmermann in A.H.* 6000 ! ; Ulanga District : Mahenge, Apr. 1932 (♀ fl.), *Schlieben* 2060 ! ; Lindi District : Mt. Kitulo, May 1903 (♀ fl.), *Busse* 2438 !

DISTR. **T**3, 6, 8; known only from Tanganyika

HAB. Lowland rain-forest and near rock-outcrops, 300–1000 m.

SYN. *C. nigrescens* Diels in E.P. IV. 94 : 296 (1910) ; T.T.C.L. 326 (1949)

var. **cardiophylla** *Troupin* in B.J.B.B. 25 : 141 (1955). Type : Tanganyika, Lindi District : Lake Lutamba, *Schlieben* 5913b (BR, holo. !, BM, S, iso !)

Leaves clearly cordate at base, usually deeply emarginate at apex, up to 11 cm. wide.

TANGANYIKA. Lindi District : Mandawa, Feb. 1935 (♂ fl.), *Schlieben* 5990 !
DISTR. **T8** ; not known elsewhere
HAB. Lowland rain-forest, 90–200 m.

8. **C. rigidifolia** (*Engl.*) *Diels* in E.P. IV. 94 : 303 (1910). Type : A.-E. Sudan, Equatoria Province, R. Nabambisso, *Schweinfurth* 3688 (B, holo !, K, iso. !)

Liane with sparingly to shortly pubescent branchlets. Leaves with densely hairy-pubescent petioles 1·8–5 cm. long and inserted 0·4–1 cm. from the base of the blade ; blade broadly ovate to somewhat rounded, subcordate to emarginate at base, obtuse to subacute and mucronulate at apex, 2·5–10 cm. long, 2·5–9 cm. wide, densely woolly-tomentellous or tomentellous beneath, sparingly pubescent to glabrescent above, clearly discolorous, papery to subcoriaceous ; basal nerves 5–10, palmate. Male inflorescences 8–30 cm. long ; bracts suborbicular, apiculate, densely pubescent-tomentose ; pedicels 0·5–1·5 cm. long. Male flowers with sparingly and longly hairy-pubescent sepals 1·2–1·5 (–2) mm. long and 0·5–0·8 mm. wide ; corolla cup-shaped, often incompletely connate, 0·5–1 mm. long, sparingly hairy-pubescent ; synandrium 6-locular, 0·5–1 mm. long. Female inflorescences 10–20 mm. long ; pedicels 1–1·5 mm. long. Female flowers with sepal 1–1·2 mm. long and 0·5–0·6 mm. wide ; petal 0·5–0·7 mm. long, 1–1·2 mm. wide ; carpel 0·5–0·7 mm. long, densely pubescent. Drupes 4–5 mm. long, 2·5–3 mm. wide, hairy-pubescent.

var. **rigidifolia**

Leaves tomentellous beneath, glabrous above ; stem sparingly puberulous to glabrescent.

UGANDA. West Nile District : hill west of Mt. Eti, July 1953 (♂ fl.), *Chancellor* 41 ! & Terego, July 1938 (♀ fl.), *Hazel* 640 !
TANGANYIKA. Njombe District : Lupembe, *Schlieben* 242 !
DISTR. **U1** ; **T7** ; also in Oubangui-Chari, the Sudan, Belgian Congo and Northern Rhodesia
HAB. Woodland or wooded grassland ; 700–1300 m.

SYN. *C. rigidifolia* (Engl.) Diels in E.P. IV. 94 : 303 (1910)
C. pareira L. var. *transitoria* Engl. subvar. *rigidifolia* Engl. in E.J. 26 : 395 (1899)
C. owariensis sensu Troupin in F.C.B. 2 : 249 (1951), *pro parte, non* DC.

var. **lanuginosa** *Troupin* in B.J.B.B. 25 : 141 (1955). Type : Tanganyika, Njombe District : Lupembe, *Schlieben* 1351 (P, holo. !, BM, BR, K, iso. !)

Leaves densely woolly-tomentellous and greyish-white beneath, sparingly pubescent above ; stem sparingly puberulous to glabrescent.

TANGANYIKA. Njombe District : Lupembe, Oct. 1931 (♂ fl.), *Schlieben* 1351 !
DISTR. **T7** ; not known elsewhere
HAB. Unknown

INDEX TO MENISPERMACEAE